AF614685

Bamboo - Recent Development and Application

Edited by Mustapha Asniza

Published in London, United Kingdom

Bamboo - Recent Development and Application
http://dx.doi.org/10.5772/intechopen.100866
Edited by Mustapha Asniza

Contributors
Abel Olorunnisola, Ajay Naik, Anju Singhwane, Avanish Kumar Srivastava, Guangjun Nie, Medha Mili, Mustapha Asniza, Norhafizah Saari, NurIzzaati Saharudin, Othman Sulaiman, Prasanth Nair, Rokiah Hashim, Sarika Verma, Vaishnavi Hada

First published in London, United Kingdom, 2024 by IntechOpen
IntechOpen is the global imprint of INTECHOPEN LIMITED, registered in England and Wales, registration number: 11086078, 167-169 Great Portland Street, London, W1W 5PF, United Kingdom

British Library Cataloguing-in-Publication Data
A catalogue record for this book is available from the British Library

Additional hard and PDF copies can be obtained from orders@intechopen.com

Bamboo - Recent Development and Application
Edited by Mustapha Asniza
p. cm.
Print ISBN 978-1-83768-414-4
Online ISBN 978-1-83768-415-1
eBook (PDF) ISBN 978-1-83768-416-8

For EU product safety concerns:
IN TECH d.o.o., Prolaz Marije Krucifikse Kozulić 3, 51000 Rijeka, Croatia,
info@intechopen.com or visit our website at intechopen.com.

Meet the editor

Dr. Asniza Mustapha studied bioresources, paper, and coatings technology in 2008. She obtained an MSc in Polymer Chemistry and Resin Technology in 2011 and a Ph.D. from Universiti Sains Malaysia, Malaysia, in 2017. Currently, she is a postdoctoral researcher at the School of Industrial Technology, Universiti Sains Malaysia. From 2009 to 2018, she worked in the field of bioresources, engaging in both fundamental and applied research, material science, and application development with a focus on biomass and lignocellulosic materials. Dr. Mustapha is now a research officer in the Bamboo, Rattan, and Palms Unit, Forest Products Division at the Forest Research Institute Malaysia (FRIM) in Selangor, Malaysia. Her research interests include biocomposite technology, natural fiber-reinforced polymer composite technology, nanocellulose, nano-based composites, and the production and development of value-added products from wood and non-wood materials. She is a member of the Malaysia Board of Technologies (MBOT) and the Malaysian Bamboo Society (MBS). As a junior researcher, Dr. Mustapha has authored about nineteen scientific articles. She has also published eleven book chapters and six general publications, including conference proceedings, short articles for magazines, and technical reports.

Contents

Preface

This edited volume, Bamboo – Recent Development and Application, is a collection of chapters on bamboo production and application. It highlights recent developments in bamboo research, as well as discusses the production and utilization of bamboo and bamboo composites. Chapters are written by scholars and edited by experts in the field.

Chapters in this book include:

1. "Introductory Chapter: Recent Research and Development on Bamboo"

2. "Production and Application of Cellulose, Dietary Fiber, and Nanocellulose from Bamboo Shoot"

3. "Recent Advances in Bamboo Research, Product Development and Utilisation in Nigeria"

4. "The Physical and Mechanical Properties of Compreg Laminated Bamboo Strips Lumber (LBSL) of *G. scortechinii*"

5. "Advances in Bamboo Composites for Structural Applications: A Review"

IntechOpen

Chapter 1

Introductory Chapter: Recent Research and Development on Bamboo

Mustapha Asniza

1. Introduction

In recent years, bamboo research has gained significant momentum as scientists and environmentalists recognize the incredible potential of this versatile plant. Bamboo, a member of the grass family, has long been valued for its strength, flexibility, and rapid growth. Recent developments in bamboo research have shed light on its unique properties, addressed challenges related to its processing and utilization, and opened up new avenues for sustainable innovation.

2. Recent research and development on bamboo

Recent developments in bamboo research have focused on unlocking the numerous untapped potentials of this remarkable plant. Researchers across various fields, including engineering, architecture, materials science, and sustainability, have been exploring innovative uses for bamboo and uncovering its unique properties. These advancements have the potential to revolutionize industries and contribute to sustainable development on a global scale.

One of the most notable recent developments in bamboo research is its potential as a sustainable alternative to traditional construction materials. Bamboo has an exceptional strength-to-weight ratio and natural durability, making it an ideal substitute for timber, steel, and concrete in various building applications [1]. Researchers have been investigating bamboo's structural properties and developing techniques to enhance its strength and durability, leading to the emergence of bamboo-based composite materials and engineered bamboo products [2–4].

Furthermore, bamboo is increasingly being recognized as a solution for mitigating climate change and promoting environmental sustainability [5]. As a fast-growing plant, bamboo has a high carbon sequestration capacity, absorbing more greenhouse gases compared to other trees. Its extensive root system also aids in soil erosion prevention and watershed management. Researchers have been studying the carbon storage potential of bamboo forests [6, 7] and exploring ways to integrate bamboo cultivation into reforestation and afforestation efforts.

Another exciting area of bamboo research lies in its potential as a renewable source of bioenergy. Bamboo's rapid growth and high biomass yield make it an excellent candidate for bioenergy production. Scientists are investigating the efficient conversion of bamboo biomass into biofuels, such as bioethanol and biogas, to reduce reliance on fossil fuels and mitigate greenhouse gas emissions [8, 9]. Besides, by harnessing bamboo's rapid growth and high cellulose content, scientists are developing cost-effective methods for biomass conversion [10], providing cleaner energy options, and reducing greenhouse gas emissions.

In addition to its construction, environmental, and energy-related applications, bamboo research has also delved into the development of bamboo-based products in various industries [11]. From textiles and furniture to paper and food, researchers are exploring the use of bamboo fibers, extracts, and by-products in manufacturing processes [12, 13], aiming for more sustainable and eco-friendly alternatives to conventional materials.

3. Conclusion

Recent developments in bamboo research have the potential to transform industries, contribute to sustainable development goals, and foster environmental conservation. The advancements in understanding its mechanical properties, environmental benefits, processing techniques, and energy applications have expanded the range of possibilities for bamboo utilization. With ongoing studies and collaborations among scientists, engineers, and policymakers, bamboo is increasingly being recognized as a valuable resource that can drive innovation, address climate change challenges, and promote a more sustainable future.

Author details

Mustapha Asniza
Forest Products Division, Forest Research Institute Malaysia (FRIM), Malaysia

*Address all correspondence to: asniza@frim.gov.my

References

[1] Van der Lugt P, Van den Dobbelsteen AAJF, Janssen JJA. An environmental, economic and practical assessment of bamboo as a building material for supporting structures. Construction and Building Materials. 2006;**20**(9):648-656

[2] Nkeuwa WN, Zhang J, Semple KE, Chen M, Xia Y, Dai C. Bamboo-based composites: A review on fundamentals and processes of bamboo bonding. Composites Part B: Engineering. 2022;**235**:109776

[3] Xiao Y. Engineered bamboo. In: Nonconventional and Vernacular Construction Materials. Sawston, United Kingdom: Woodhead Publishing; 2016. pp. 433-452

[4] Sharma B, van der Vegte A. Engineered bamboo for structural applications. In: Nonconventional and Vernacular Construction Materials. Sawston, United Kingdom: Woodhead Publishing; 2020. pp. 597-623

[5] Yiping L, Yanxia L, Buckingham K, Henley G, Guomo Z. Bamboo and climate change mitigation: A comparative analysis of carbon sequestration. International Network for Bamboo and Rattan. 2010;**30**:1-47

[6] Vogtländer JG, van der Velden NM, van der Lugt P. Carbon sequestration in LCA, a proposal for a new approach based on the global carbon cycle; cases on wood and on bamboo. The International Journal of Life Cycle Assessment. 2014;**19**:13-23

[7] Xu X, Xu P, Zhu J, Li H, Xiong Z. Bamboo construction materials: Carbon storage and potential to reduce associated CO_2 emissions. Science of the Total Environment. 2022;**814**:152697

[8] Chin KL, Ibrahim S, Hakeem KR, San H'ng P, Lee SH, Lila MAM. Bioenergy production from bamboo: Potential source from Malaysia's perspective. BioResources. 2017;**12**(3):6844-6867

[9] Hada S, Roat P, Chechani B, Kumar S, Yadav DK, Kumari N. An overview on biomass of bamboo as a source of bioenergy. In: Biotechnology for Biofuels: A Sustainable Green Energy Solution. Singapore: Springer; 2020. pp. 241-265

[10] Ding Z, Awasthi SK, Kumar M, Kumar V, Dregulo AM, Yadav V, et al. A thermo-chemical and biotechnological approaches for bamboo waste recycling and conversion to value added product: Towards a zero-waste biorefinery and circular bioeconomy. Fuel. 2023;**333**:126469

[11] Borowski PF, Patuk I, Bandala ER. Innovative industrial use of bamboo as key "green" material. Sustainability. 2022;**14**(4):1955

[12] Hakeem KR, Ibrahim S, Ibrahim FH, Tombuloglu H. Bamboo biomass: Various studies and potential applications for value-added products. Agricultural Biomass Based Potential Materials. Cham, Switzerland: Springer; 2015. pp. 231-243

[13] Silva MF, Menis-Henrique ME, Felisberto MH, Goldbeck R, Clerici MT. Bamboo as an eco-friendly material for food and biotechnology industries. Current Opinion in Food Science. 2020;**33**:124-130

Chapter 2

Production and Application of Cellulose, Dietary Fiber, and Nanocellulose from Bamboo Shoot

Guangjun Nie

Abstract

The cellulose from bamboo has excellent toughness, hygroscopicity, and high crystallinity. Bamboo shoot dietary fiber can modulate the gut microbiota to prevent high-fat diet-induced obesity and can be applied for food fortification. Bamboo shoot contains a low content of lignin and is extracted easily for nanocellulose, which is used to prepare all kinds of composite materials. In this chapter, lignification process of bamboo shoot shells will first be discussed to reveal the principle of lignification. Then, the preparation methods and applications of cellulose, dietary fiber, and nanocellulose from bamboo shoots that were successively generalized to further improve the exploration and application of bamboo shoots or bamboo shoot wastes such as bamboo shoot shells.

Keywords: bamboo shoots, cellulose, dietary fiber, nanocellulose, production, application

1. Introduction

Bamboo shoots (BSs) have long been used as food and in traditional medicine in many Asian countries because the tender BSs are rich in nutrients and bioactive compounds associated with health benefits against many chronic and degenerative diseases [1]. For example, bamboo shoot powder supplementation mainly consists of cellulose, hemicellulose, and lignin, which can reduce the levels of triglycerides, blood glucose, total cholesterol, high-density lipoprotein, and low-density lipoprotein in mice, and decrease the risk of fatty liver disease [2]. However, up to 70% of the harvested BSs have been discarded as waste biomass in China and other parts of the world. Therefore, it is of great importance for the exploration and application of BSs to develop the technologies and methods for novel products of BSs.

Presently, BSs and the processing residue have been studied as cheap sources of valuable nutrients and bioactive compounds for value-added products (e.g., food additives, functional foods, and pharmaceuticals) [1]. Dietary fiber (DF), defined as the seventh nutrient for humans by nutritionists, has many benefits for our health, such as improving the intestinal flora and decreasing the probability of obesity and cardiovascular disease [3, 4]. The total dietary fiber (TDF) in BSs ranges from 2.23 to 4.20 g/100 g fresh weight of shoots [5]. The DF from BSs increases food quality as well as organoleptic properties [5]. The cellulose from BSs has been described as having excellent toughness, hygroscopicity, and high crystallinity [6], and can be regarded as an ideal material for preparing BSs cellulose nanocrystals (BSCNC). Cellulose nanocrystal (CNC) has many interesting structural features and unique physicochemical properties, including magnificent mechanical strength, high surface area, and many hydroxyl groups for chemical modification, low density, and biodegradability. It has attracted great interest from researchers in such fields as biomedicine, electronic gadget, water purification, nanocomposite (NC), and membrane [7–9]. BSs have many advantages such as low content of lignin, fast growth, and wide distribution, which represent a favorable scenario for producing CNCs from BSs with low chemical and energy consumption [10]. Therefore, it is vital for the exploitation and utilization of BSs to produce BSC, BSDF, and BSCNC. Besides, the degree of polymerization and crystallinity of cellulose increased with the maturity of BSs, which positively correlates with the lignification [11]. Thus, an understanding of lignification process of BSs is helpful for the production of BSC, BSDF, and BSCNC.

2. Lignification process of bamboo shoot

Lignin is recognized as a limiting factor to biomass-to-products conversion, and thus it is required for efficient exploitation of BSs to understand the molecular mechanisms underlying lignin deposition. In general, the lignification of BSs takes place in the two stages of growth and storage.

During the growth of BSs, the lignification degree and lignin content increased with the height of BSs [12], and the lignin deposition increased with internode maturity and from bottom to top in each internode, except for the mature internode (8 m), which showed higher lignin levels in the lower portion as shown **Figure 1a**. In this stage, lignification is attributed to secondary cell wall (SCW) formation. The biosynthetic pathways are highly regulated by transcription factors (TFs). The network of TFs is mainly divided into two categories: those in the MYB family and those in the NAC family. NAC TFs act as "master switches" and regulate downstream levels of TFs to initiate SCW synthesis. MYB TFs play a central role in the transcriptional regulation of the deposition of plant SCW materials and have been reported to function as a link between upstream NAC TFs and downstream structural genes [14]. The lignification degree positively correlates with the content of lignin and cellulose. MYB20, MYB42, MYB43, and MYB63 may be positive regulators of both lignin and cellulose

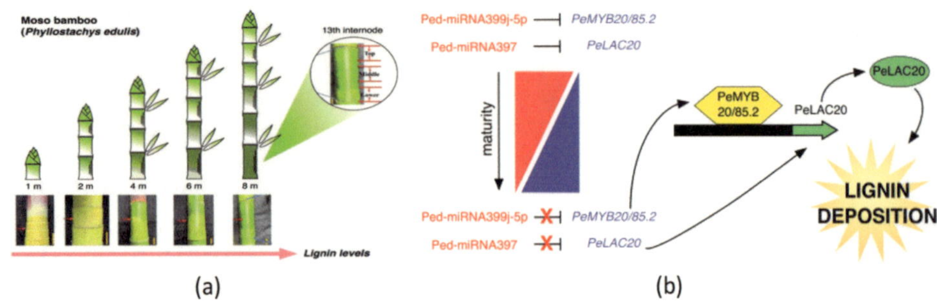

Figure 1.

Lignification process of BSs (P. edulis) (a) and miRNA-mediated "MYB-PeLAC20" regulatory module for lignin polymerization of BS (b) [12, 13]. Internodes were also divided into three portions, named lower, middle, and top.

biosynthesis in *Phyllostachys prominens* shoots (HBSes), and MYB20 and MYB43 had a positive correlation with the lignin content in *Phyllostachys edulis* shoots (MBSes) [15]. It is commonly recognized that MYB20 plays an important regulatory role in the lignin biosynthesis pathway. As schemed in **Figure 1b**, the lignin polymerization of BSs based on the regulation of MYB20 was proposed: (1) in young internodes, higher expression of Ped-miRNA399j-5p and Ped-miRNA397 represses the expression of PeMYB20/85.2 and PeLAC20; (2) the expression levels of Ped-miRNA399j-5p and Ped-miRNA397 decrease with the maturity of BSs, resulting in the up-regulation of PeMYB20/85.2 and PeLAC20; (3) PeMYB20/85.2 binds to the promoter of PeLAC20, boosting its expression and inducing lignin deposition [12, 13].

Lignification has a close correlation with the relevant key enzymes in the growth of BSs. The phenylalanine ammonia-lyase (PAL) and peroxidase (POD) activities were progressively reduced from shoot base to top, and the contents of cellulose and lignin also decreased [16]. It follows that PAL and POD contribute to the lignification. Four key enzymes in lignin cell wall deposition are PAL, cinnamyl alcohol dehydrogenase (CAD), POD, and laccase. PAL is the first rate-limiting enzyme. CAD degrades hydroxy-cinnamaldehydes into the corresponding alcohols. POD and laccase catalyze the polymerization of cinnamyl alcohols into lignin in the last step of lignin synthesis. Accordingly, the accumulation of lignin in fresh tissue was positively correlated to the increasing activity of PAL, CAD, POD, and laccase [17, 18]. However, PAL and laccase activities first increased and then decreased with the growth of BSs [12]. The crystalline structure of cellulose fibers is surrounded by hemicellulose and lignin. Lignin is a heterogeneous aromatic polymer found as 10–35% of lignocellulose. The resistance of lignin to breakdown is a major obstacle in the bioconversion of lignocellulose. The microbial breakdown of lignin attracts considerable interest. White-rot fungi are known to break down lignin with the aid of extracellular POD and laccase enzymes [19]. Coculturing microorganisms (*Pleurotus ostreatus* and *Aspergillus niger*) have been adapted to decompose the lignocellulose of bamboo shoot shells (BSSs). During the degradation, there was a rise in the activities of enzymes carboxymethylcellulase (CMCase) and laccase [20].

After BSs are harvested, the storage is the second stage of BSs for lignification, which is also correlated with the expression of key enzymes such as POD, PAL, and CAD [18]. The contents of cellulose and lignin increased rapidly with storage time. In the first five days, cellulose increased rapidly, and in the next five days, it became to slow down apparently. However, the lignin content increased at an even speed during the storage time. Meanwhile, the activities of POD and PAL increased significantly with the storage time [18, 21]. The lignification during postharvest storage of BSs may be correlated with the up-regulated expression of the IFs of SND2, KNAT7, MYB20, and MYB85 from the NAC and MYB families. Application of melatonin effectively retarded the lignification, as well as reduced lignin and cellulose contents by inhibiting the activities of PAL and POD, but enhancing those of superoxide dismutase, catalase, and ascorbate peroxidase activities. This was mainly attributed to the inhibition of the TFs [14]. Therefore, lignification has been considered a typical characteristic of senescence of BSs during postharvest storage, which is affected by various factors to name a few. Cold storage can reduce weight loss, browning, respiration rates, and sugar degradation in BSs; decrease related enzymatic activities; and inhibit the increase in lignin and cellulose content. The increase in the degree of lignification and fibrosis is the main reason for senescence and for the decline in quality of BSs after harvest [22]. Low temperature (4°C) could decrease the activities of POD and PAL significantly, as well as decrease the contents of cellulose and lignin significantly [21].

However, the firmness, and lignin and cellulose content of BSs increased and accelerated by higher storage temperature. The increase in firmness of BSs during storage is a consequence of tissue lignification, a process associated with increase in the activities of PAL, CAD, and POD [18]. Hydrogen peroxide (H_2O_2) accelerates the programmed cell death process by upregulating DNase, RNase, and caspase 3-like activities and enhances the lignification of BSs. Therefore, endogenous H_2O_2 may play a vital role in the lignification process of BSs [23]. On the other hand, lignin biosynthesis has three stages: biosynthesis, transport, and polymerization of its precursors. Wherein, the transport activities for coniferin and β-glucocoumaryl alcohol depend on vacuolar type H^+-ATPase and a H^+ gradient across the membrane. Thus, the transportation is mediated by secondary active transporters energized partly by the vacuolar-type H^+-ATPase [24].

3. Preparation and application of BSC

Like wood, the cell walls of bamboo culms consist mainly of cellulose, hemicellulose, and lignin. Bamboo cellulose has excellent toughness, hygroscopicity, and high crystallinity [6]. Cellobiose is the basic repeating unit of cellulose. In the molecular structure of the cellobiose, strong intramolecular hydrogen bonds among the contiguous glucose segments in the chain and intermolecular hydrogen bonds with various other surrounding chains are formed, which are responsible for tight packing of crystalline regions of cellulose fibrils (**Figure 2a**). As shown in **Figure 2b**, the extensive orientation of the hydrogen bonding network and different possible directions of glucose units lead to various crystalline forms for cellulose, categorized into four types (Cellulose I, II, III, and IV) [26, 27]. Cellulose I has a parallel packing in the hydrogen-bonded cellulosic network, which is the typical crystalline form of native cellulose. On the contrary, cellulose II has an antiparallel packing of the hydrogen-bonded cellulosic network, which is either found in chemical recovery of cellulose I by solvating in some solvents (acid or base). Modification of Cellulose I & II using ammonia leads to cellulose III, whereas cellulose IV can be derived after heating cellulose III in glycerol at 260°C [28].

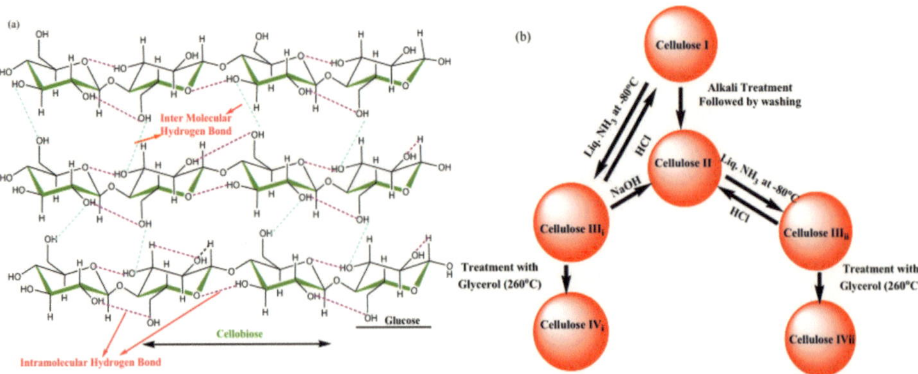

Figure 2.
Possible inter as well as intramolecular hydrogen bonding between different cellulose units (a) [25] and different allomorphs of cellulose (b) [26, 27].

As listed in **Table 1**, BSC has successfully been extracted from bamboo plants by chemical, physical, biological, and combinational methods to name a few. The hydrothermal treatment with NaOH and $NaClO_2$ has been found an effective method to prepare BSC, and these chemical reagents aim to remove hemicellulose and lignin of BSSF, and then BSSF was treated with alkali and bleached to produce BSC [34]. To protect environment, the environmental protection oxidant sodium percarbonate combined with alkali oxygen bath method was used to prepare BSSF. The method had a good effect on the degumming of BSSF, and most of the noncellulose components such as hemicellulose, lignin, and pectin had been removed. The degumming rate was 69.17% at the scouring temperature of 90°C, the alkali boiling time was 120 min, the dosage of sodium percarbonate was 24 g/L, and the dosage of H_2O_2 was 40 mL/L. The cellulose content and crystallinity were 73.19 and 61.40%, and the fiber had a smooth surface and a typical cellulose structure with better thermal stability [30]. The physical ultrasound-assisted method has often been coupled with some chemical methods to improve the preparation of BSC. The optimum parameters of alkali oxygen bath degumming were shown as follows: Scouring temperature of 95°C, alkali boiling time of 120 minutes, sodium hydroxide dosage of 20 g/L, and H_2O_2 dosage of 24 mL/L. The corresponding degumming rate was 70.07%. Pectin, lignin, and hemicelluloses were effectively removed from BSSF. The extracted BSSF presented a smooth surface, good quality, and high cellulose content, exhibiting a remarkable degumming effect of the ultrasonic pretreatment. The addition of a certain amount of sodium hydroxide in the ultrasonication process enhanced the "hollow effect" of the ultrasonic waves, improving the degumming effect [31]. The combinational use of acid and alkaline is more beneficial for BSC preparation than single use. As shown in **Figure 3**, acid pretreatment combined with alkali-oxygen bath degumming process removed most of pectin and non-cellulosic contents, and the prepared BSSF has a high content of 73.19% cellulose, exhibiting an excellent quality. The degummed fibers were separated from each other in a single fiber state, showing typical cellulose I structure, and with a large number of grooves on the surface [29]. By comparison, the acidic EtOH-HNO_3 process efficiently removed hemicellulose and lignin from bamboo, achieving

Method	Traits of BSC	Ref.
Acid pretreatment combined with alkali-oxygen bath degumming process	The crystallinity increased from the control (43.53%) to final fiber (64.35%). 73.19% of cellulose content.	[29]
Oxidant sodium percarbonate combined with alkali oxygen bath method	The cellulose content and crystallinity were 73.19 and 61.40%. The degumming rate was 69.17%.	[30]
Ultrasound-assisted alkali-oxygen bath method	Degumming rate was 70.07%. High cellulose content, smooth surface, and good quality.	[31]
Enzyme (cellulase and hemicellulose) coupled with LAB fermentation	The crystallinity and the thermal stability were improved.	[32]
One-step process (EtOH-HNO_3)	The highest purity of cellulose (96.8%) with the highest degree of polymerization (DP) value (815).	[33]

Table 1.
Preparation method and traits of BSC.

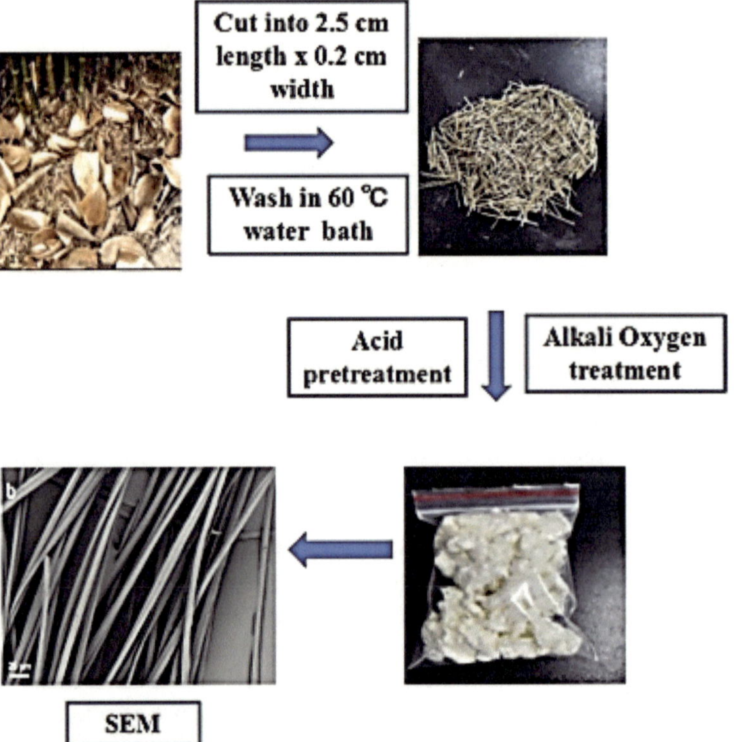

Figure 3.
Degumming process of BSSF [29].

the highest purity of cellulose (96.8%) with a high degree of polymerization (815) [33]. Bamboo shoot shells (BSS) were used as feedstock for the production of BSSF. Biological method is an effective way to improve the effect of BSS silage. Cellulase, hemicellulase, and lactic acid bacteria (LAB) can reduce the silage pH but increase the concentrations of dry matter and protein of BSS. The ensilaged fiber had an increased crystallinity and thermal stability [32].

BSC has been mainly applied as an ideal raw material in the synthesis of high-value-added material. Except for being used to prepare BSCNC, BSC, as a matrix, has often been explored to prepare many composites to name a few. BSSF was blended with starch/poly(lactic acid) (PLA) matrices to create a new low-cost biodegradable ternary composite. The mechanical strength, surface wettability, and water absorption of the composite have been increasingly improved with the rise of BSSF content from 0 to 20 wt% [35]. Also, it was used to cross-link with sodium alginate (SA) based on hydrogen bonding. The composite aerogels exhibited better encapsulation efficiency and *in vitro* antioxidant activity for the encapsulation and sustained release of bioactive compounds [36]. Besides, the modified BSC can exhibit some novel traits. The crude cellulose isolated from bamboo shoot processing by-products (BPSs) can be esterified using soybean oil to form esterified microcrystalline cellulose (E-MCC). The E-MCC had better dispersion and compatibility in starch film matrix. 5 wt% E-MCC can enhance the tensile strength and Young's modulus of TPS [37].

4. Preparation and application of BSDF

The major health benefits of DF are proper and regular bowel movement, prevention of constipation, diarrhea, diverticulitis, hemorrhoids, cardiac diseases, and cancer. Presently, many food products are being fortified with DF to provide an adequate amount in the diet. DF from BSs increases food quality as well as organoleptic properties [5]. DF is basically nondigestible and nonabsorptive parts of plants. DF is classified as insoluble dietary fiber (IDF) and soluble dietary fiber (SDF). From the same species, IDF showed better adsorption capacity than SDF. This is ascribed to that the surface of IDF is porous, whereas the SDF is relatively flat and compact, and stronger than IDF in the *Lactobacillus* and *Bifidobacterium* promotion effects [38]. As a cheap potential DF resource, fresh BS has a high content of TDF ranging from 2.23–4.20% [5]. Some methods to extract DF from BSs have been presented in **Table 2**, including physical method, chemical method, biological method, and combined method. Physical method is the most commonly used method to extract DF from BSs. High-temperature cooking (HTC), high-pressure homogenization (HPH), ultrasonic treatment (UT), and their combination (i.e., UT–HTC, HPH–HTC, and UT–HPH) have been applied to modify BSDF. The SDF content increased in all samples. UT–HPH could be a promising and alternative modification method to obtain high-quality BSDF. The modifications can destroy the hydrogen bonds between the lignin and hemicellulose of BSDF, significantly decreased the particle size of BSDF, and increased the SDF content, which increased oil holding capacity (OHC) and water swelling capacity (WSC) in BSDF by 55.35 and 91.47%, respectively. It is considered that combined modification technologies are effective methods to improve the quality of BSDF [42].

Chemical method is also another important method to extract DF from BSs. The acid-alkali chemical method was used to prepare BSDF, which had higher TDF with a higher crystalline region [39]. Enzymolysis can improve the DF in the solubility by decomposing IDF, and producing SDF [39, 46]. With BSS as substrate, the extraction yields of IDF and SDF were 56.21 and 8.67%, respectively. The resulting fibers showed significant WSC, WHC, and *in vitro* binding capacities to fat, cholesterol, bile acids, and nitrites [46]. No-matter-chemical method and enzymolysis both improve functional properties of DF, which exhibits many good behaviors such as water retention capacity (WRC) (11.24–15.13 g/g), WSC (18.84–28.75 mL/g), OHC (6.71–10.15 g/g), glucose adsorption capacity (GAC) (0.08–6.89 mmol/g), and glucose retardation index (GRI) (3.57–40.92%). Thus, the functionalities of DF can be improved by manipulating its physic-chemical properties [39]. An effective solvent can also improve the extraction of DFs from BSs. Subcritical water (SW) and high-pressure homogenization (HPH) was used to treat BSs for the production of DF. The functions and the structural characteristics of DFs from BSs were significantly improved, and the SDF content was dramatically increased. SW modification enhanced WHC, OHC, and WSC of DFs better than HPH. The modified DF had increasing abilities to absorb cholesterol and nitrite ions [40].

The functional properties of DF change with the variation of extraction method. Alkali extraction (AE), enzymatic extraction (EE), ultrasonic-assisted enzymatic extraction (UAEE), and shear homogeneous-assisted enzymatic extraction (SHAEE) were applied to extract BSDF. The extracted BSDF by AE had the lowest protein content and crystallinity index, but the extracted BSDF by EE had highest protein

Method	Traits of BSDF	Ref.
Acid-alkali method	Higher TDF and higher SDF, WRC (11.24–15.13 g/g), WSC (18.84–28.75 mL/g), OAC (6.71–10.15 g/g), GAC (0.08–6.89 mmol/g), and GRI (3.57–40.92%).	[39]
Modified AOAC enzymatic-gravimetric method		
SW	The crystallinity decreased, but SDF content and the ability to absorb cholesterol and nitrite ions increased.	[40]
HPH		
Lactobacillus and *Bifidobacterium* fermentation	Strong cholesterol-adsorption activity and prebiotic potential.	[38]
AE	The BSDF with the lowest protein content and crystallinity index.	[41]
EE	EE: The BSDF with the lowest OHC and highest protein content.	
UAEE	UAEE: The BSDF with the highest OHC and GAC	
SHAEE	SHAEE: The BSDF with the highest SDF content, WHC, α-amylase activity inhibition ratio, and the smallest particle size, and with a porous and loose structure. Hypoglycemic activity of the four BSDF samples generally followed the order of SHAEE > UAEE > EE > AE	
HPH	The smallest particle size of BSDF	[42]
UT	A loose structure with a honeycomb appearance on the surface.	
Combinations (i.e., UT–HTC, HPH–HTC, and UT–HPH)	The lowest relative crystallinity. UT-HPH: The particle size decreased, but the SDF content increased. UT–HTC: SDF has a more porous, looser morphological structure, and increased content.	
LGG fermentation	The structure and composition of TDF changed, the proportion of IDF increased, and the TDF was decomposed into small pieces. Great WSC, stronger NIAC, the ability to produce SCFAs, and stronger anti-digestion ability.	[43]
Enzymatic hydrolysis	Particle sizes and the microstructure of IDF greatly changed.	[44]
DHPM	A honey-comb appearance and large cavities on the surface of modified fiber, which has increased crystallinity and thermal stability, higher WHC, and more promising binding capacities for oil, nitrite ion, glucose, and cholesterol.	
ME	The yields of IDF and SDF were 56.21 and 8.67%. The fibers showed significant WSC, WHC, and binding capacities to fat, cholesterol, bile acids, and nitrites.	[45]

Table 2.
Preparation method, traits, and applications of BSDF.

content and the lowest OHC. The BSDF generated by UAEE had the highest OHC and glucose adsorption capacity (GAC). SHAEE obtained the highest SDF content (17.89%), WHC (8.81 g/g), and α-amylase activity inhibition ratio (19.89%) and the smallest particle size (351.33 μm). BSDF extracted by SHAEE and UAEE presented a porous and loose structure. In terms of *in vitro* hypoglycemic activity, the four

BSDF samples generally followed the order of SHAEE>UAEE>EE > AE. SHAEE is an innovative and promising method to obtain BSDF with excellent physicochemical and functional properties [41]. Therefore, the combination use of various methods is an effective road to change the functional properties of BSDF.

Microbial fermentation is composed of multiple enzymatic reactions. In view of the effect of enzymatically method on BSDF extraction, microbial fermentation ought to have an apparent effect on BSDF preparation. *Lactobacillus rhamnosus* GG (LGG) fermentation can change the structure and composition of TDF of BSs. The TDF was decomposed into small pieces, and the proportion of IDF increased [43]. The enzymatically SDF exhibited significantly higher glucose adsorbing capacity than those of IDF and TDF and showed similar inhibition potential against α-amylase with acarbose. TDF displayed the greater capacities for delaying glucose diffusion and inhibition of α-glucosidase than those of SDF [46]. The structure and properties of BSIDF can be changed for extensive applications by some modifications of enzymatic hydrolysis and dynamic high-pressure micro-fluidization (DHPM). The particle sizes of IDF powders significantly decreased, and the microstructure made marked changes. The treatments removed part of lignin and hemicellulose and increased the crystallinity and thermal stability of the modified fibers. Relative to the modification of enzymatic hydrolysis, the modification of DHPM better decreased BSIDF in particle size and porous structure, and the DHPM-modified BSIDF exhibited not only higher WHC but also more promising binding capacities to oil, nitrite ion, glucose, and cholesterol. It follows that DHPM could more effectively improve functional properties of BSIDF than enzymatic hydrolysis [44]. On the other hand, BSIDF was prepared and used as plant food particle stabilizer for oil-in-water (O/W) Pickering emulsions. The DFs from bamboo shoots had a soft nature and suitable shape to produce stable Pickering emulsions, which could be used as food-grade particles for applications in food and cosmetics industries. The BSDF suspensions and BSDF-stabilized O/W emulsions both exhibited shear-thinning behaviors; moreover, both viscosity and module were increased with the increase of BSDF contents. The surface coverage of emulsions was positively correlated with the content of BSDF suspensions. The obtained Pickering emulsions would have a wide application in food and cosmetics field [47].

BSDF has been mainly applied in food industry, especially in the field of healthy diet. BSDF has been shown to prevent high-fat diet-induced obesity through modulating the gut microbiota, which is a potential prebiotic fiber that modulates the gut microbiota and improves host metabolism and is the most effective in suppressing high-fat diet-induced obesity [48]. The administration of BSDF improved the lipid metabolism disorderly situation of hyperlipidemia mice. Compared with normal group, TDF supplement could exhibit the lowest body weight gain (2.84%) in mice [46]. Several potential mechanisms have been suggested to be responsible for the hypolipidemic effects of BSDF. (1) The fibers are not able to be digested in the gastrointestinal tract, resulting in greater satiety and less calorie intake. (2) The fibers present in gastrointestinal tract disturb lipid absorption due to their binding capacities. (3) The binding or adsorptive capacities of bile acids and cholesterol enhance their removal from enterohepatic circulation, leading to lower cholesterol in the various pools. In addition, the fibers might effectively regulate the transcription of certain genes involved in lipid catabolism and modulate the gut microbiota to improve lipid profiles. These mechanisms might work together to control the body weight gain and improve the lipid metabolism disorderly situation [45]. The administration of BSDF decreases total cholesterol, triglyceride, and low-density lipoprotein-cholesterol by 31.53, 21.35, and 31.53%, respectively [46], exhibiting a good potential as a cholesterol-lowering ingredient or an adjuvant for functional food with anti-obesity and hypolipidemic effects [49]. Besides, BSDF can be used to enhance the pasting viscosity and viscoelasticity of rice starch. BSDF was coated on starch granules to form a protective layer that restricts the accessibility of hydrolytic enzymes to starch and reduces the extent and rate of starch digestion, improving the properties of starch and facilitating its utilization [50]. BSIDF could be used to formulate desirable physicochemical properties of starchy foods. High content and appropriate size of BIDF could cause rice starch granules to aggregate *via* strong intra- and intermolecular hydrogen bonds to form a tight honeycomb-like and dense porous structure. As a result, peak viscosity, final viscosity, hardness, and the long-term retrogradation of rice starch paste increased [51]. BSDF can improve bread in texture properties except for hardness, and had good ability to absorb cholesterol and anti-digestion properties [43].

5. Preparation and application of BSCNC

CNCs have many interesting structural features and unique physicochemical properties, including magnificent mechanical strength, high surface area, and many hydroxyl groups for chemical modification, low density, and biodegradability, and thus have attracted great interest from researchers in various fields of biomedical, electronic gadgets, water purifications, NC, and membranes [7–9]. A persistent progression is going on in the extraction, modification, and application of CNCs as shown in **Figure 4**. BSs have many advantages such as low content of lignin, fast growth, and wide distribution, which represent a favorable scenario for producing CNCs from BSs in low chemical and energy consumption [10]. As indicated in **Figure 5**, the degree of polymerization and crystallinity of cellulose increased with the degree of maturity of the raw bamboo samples. The cellulose nanofiber (CNF) can be obtained from immature bamboo, which

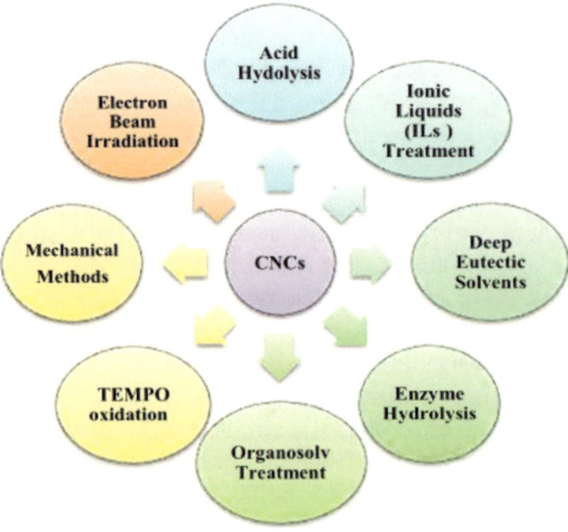

Figure 4.
CNCs: Methodologies for CNCs preparation [7].

contains <10 wt% of lignin by mild mechanical treatment. Immature bamboo can be a very promising source of raw material for the production of CNF [11]. Therefore, a biomass with less lignin may be advantageous to the production of high-performance CNF because the delignification step could be omitted from the

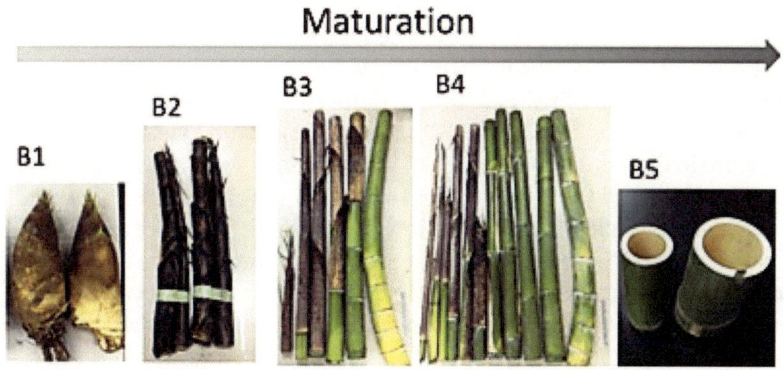

Figure 5.
Pictures showing bamboo samples under different growth stages [11].

Method	Traits of BSCNC	Application	Ref.
Acid hydrolysis	86.96% crystallinity and 22% yield of CNCs.	Biomedical and food packaging application	[54]
Combinational method (low content of sulfuric acid hydrolysis/ ultrasonic treatment)	Both CNF and CNCs displayed high crystallinity indexes of 68.51 and 78.87%.	Functional components or carriers in food and pharmaceutical fields.	[55]
TEMPO-mediated oxidation method	Good mechanical property and dimensional stability, 60–90 nm in the diameter of BSNC. The crystallinity increased, but the thermal stability decreased.	Filter and adsorption materials.	[56]
Hydrothermal treatment with NaOH and $NaClO_2$.	The least stability, its better crystallinity.	Paint formulation and nanocarriers in biomedical fields.	[34]
Alkali and hydrogen peroxide pretreatment	The yield of 50.67% with a crystals recovery of 77.99%		[9]
Water, bleaching, and alkali treatment	5.80 to 8.57 nm in diameter and 82.93 to 170.67 nm in length for BSNC materials.	Reinforcing PVA Films	[10]
Chemical and mechanical separate pre-treatment prior to sulfuric acid hydrolysis	Average widths of 134.2 + 34.33 nm. Crystallinity degrees of the nanofibrils and crystals were 49.47 and 56.92%.	Materials	[53]

Table 3.
Preparation method, traits, and applications of BSCNC.

process [52]. However, mature bamboos contain more cellulose, which contributes to the isolation of CNCs and CNF with the use of different chemical and mechanical methods [6, 53].

As listed in **Table 3**, there are many methods to prepare CNCs from BSs. Acid hydrolysis is one of the traditional and universally used methods for the preparation of CNCs. Several mineral acids, for example, sulfuric acid (H_2SO_4), muriatic acid (HCl), hydrobromic acid, and orthophosphoric acid (H_3PO_4) were utilized for the preparation of nanocrystals [7, 10]. The mechanism for the formation of CNCs is schemed as in **Figure 6**. Non-crystalline regions of cellulose are removed by acid hydrolysis. As BSs were treated using 50% v/v H_2SO_4 at 50°C for 45 min, the resulting CNCs had 5.80 to 8.57 nm in diameter, 82.93 to 170.67 nm in length, and 11.89 to 21.97 in aspect ratio [10]. The extracted CNC by acid hydrolysis has a higher degree of crystallinity of 86.96% and a higher yield of 22% [54]. However, sulfuric acid hydrolysis decreased the thermal stability of CNCs [55], the extracted retained some lignin, and thus the chemical pretreatment is necessary to remove the purities prior to acid hydrolysis [53].

A combinational method has been proven to be better for the extraction of BSCNC than a single-acid hydrolysis method [55], and different combinations can lead to various effects to name a few. As BS was complexly treated by acid hydrolysis and the pretreatment using alkali and hydrogen peroxide, the optimum CNC yield of 50.67% with a crystals recovery of 77.99% was obtained at the sulfuric acid concentration of 54.73% and a temperature of 39°C. The sulfuric acid concentration has a more significant effect than the temperature. The crystals recovery of CNC higher than 70% was obtained at a sulfuric acid concentration of around 55–60%. The temperature has no significant effect on the crystals recovery of CNCs [9]. The dosage of H_2SO_4 is very high, which increases environmental burden. Ultrasonic treatment can reduce the dosage of acid. The combination of low-concentration acid hydrolysis and ultrasonic treatment was used to extract CNF and CNCs from BSs (*Leleba oldhami Nakal*). NFC and CNCs exhibited typical long-chain and needle-like structures, respectively.

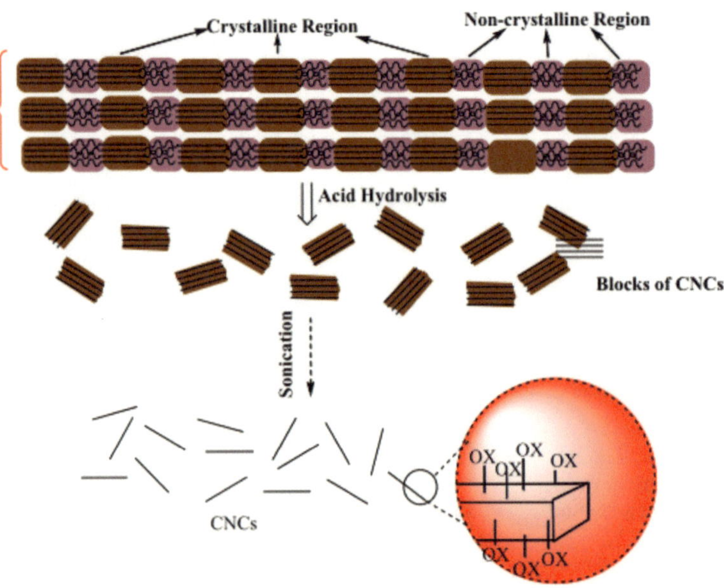

Figure 6.
Schematic representation of CNC formation from cellulose acid hydrolysis [25].

Lignin and hemicellulose were successfully removed, and both CNF and CNCs displayed high crystallinity indexes of 68.51 and 78.87% [55].

BSCNC can be potentially used in various fields. The CNCs prepared by acid hydrolysis has a high crystallinity value and a higher capability to enhance mechanical properties of polymeric composites, and thus is potentially applied in NC for biomedical and food packaging application [54]. The well-dispersed BSNC showed good reinforcement effects on the PVA matrix, and the BSNC-reinforced PVA films exhibited a tensile strength of 32.46 MPa, 53% higher than that of neat PVA film [10]. Besides, BSCNC has gained increasing interest due to its excellent properties and great potential as a functional component or carrier in food and pharmaceutical industries [55]. BSCNC showed great potential as an emulsifier in the Pickering emulsions [57]. With the increasing BSNC content, the emulsions presented increased droplet size, and even demulsification occurred. The surface coverage was above 100% for the Pickering emulsions. All emulsions showed elastic behaviors.

The modification of BSSF extends the application field. As raw material, BSSF was used to prepare self-reinforced NC composite by a simple method mediated with 2,2,6,6-tetramethylpiperidine-1-oxy (TEMPO) oxidation. Compared with the original fiber, the crystallinity of the composites increased, while the thermal stability decreased. The nanocomposite has good dimensional stability indicating the possibility for replacing hard-wood resources and the great potential for utilizing agricultural wastes. It provides a promising and convenient route to obtain film sheet materials with micro- or nano-structures from nature cellulose fibers [56]. Due to their hydrophilic tendencies, CNCs are modified to act as hydrophobic drug carriers. Rarasaponins (RSs) were attached onto BSCNCs to enhance their hydrophobicity. The curcumin uptake on CNCs-RSs reached 12.40 ± 0.24%, but it was slowly released until approximately 78% in three days [8].

6. Conclusion

There are increasing studies focused on the preparation methods and applications of cellulose, DF, and CNCs from BSs. The methods mainly include physical, chemical, biological, and combinations. As compared with every single method, the combinations exhibit more significant effect on the preparation and modification of BSC, BSDF, and BSCNC. With increasing emphasis on environmental protection, there will be increasing attention to novel technical processes to produce the products from BSs in order to efficient and green exploration and application of BSs. Due to various traits to be developed and applied, the extracted and/or modified BSDF have presently been applied as functional foods or food additives in food industry, and the extracted and/or modified BSCNC are used as materials in food, biomedical, pharmaceutical fields, and so on. In the future, novel modifications will continue to be created to discover some new traits of BSC, BSDF, and BSCNC, which will be developed into more new products that people need.

Acknowledgements

This research work was supported by Natural Science Foundation of Anhui Province (2108085MC71) and State Key Laboratory of Microbial Technology Open Projects Fund (Project NO. M2022-18).

Author details

Guangjun Nie
College of Biological and Food Engineering, Anhui Polytechnic University, Wuhu, China

*Address all correspondence to: ngjason@ahpu.edu.cn

References

[1] Lin Z, Chen J, Zhang J, Brooks MS-L. Potential for Value-Added Utilization of Bamboo Shoot Processing Waste—Recommendations for a Biorefinery Approach. Food and Bioprocess Technology. 2018;**11**(5):901-912

[2] Yang J, Wu L, Yang H, Pan Y. Using the Major Components (Cellulose, Hemicellulose, and Lignin) of Phyllostachys praecox Bamboo Shoot as Dietary Fiber. Frontiers in Bioengineering and Biotechnology. 2021a;**9**:669136

[3] Gan J, Xie L, Peng G, Xie J, Chen Y, Yu Q. Systematic review on modification methods of dietary fiber. Food Hydrocolloids. 2021;**119**:106872

[4] Nirmala C, Bisht MS, Laishram M. Bioactive compounds in bamboo shoots: health benefits and prospects for developing functional foods. International Journal of Food Science and Technology. 2014;**49**(6):1425-1431

[5] Bishta MS, Nirmalab C, Santoshb O. Bamboo Shoot as a Potential Source of Dietary Fiber for Food fortification. Food and Pharmaceuticals. 11th World Bamboo Congress Mexico. 1 Aug 2018

[6] Yu M, Yang R, Huang ACL, Cao X, Yang F, Liu D. Preparation and characterization of bamboo nanocrystalline cellulose. BioResources. 2012;7(2):1802-1812

[7] Rana AK, Frollini E, Thakur VK. Cellulose nanocrystals: Pretreatments, preparation strategies, and surface functionalization. International Journal of Biological Macromolecules. 2021;**182**:1554-1581

[8] Wijaya CJ, Ismadji S, Aparamarta HW, Gunawan S. Hydrophobic Modification of Cellulose Nanocrystals from Bamboo Shoots Using Rarasaponins. ACS Omega. 2020;**5**(33):20967-20975

[9] Wijaya CJ, Ismadji S, Aparamarta HW, Gunawan S. Optimization of cellulose nanocrystals from bamboo shoots using Response Surface Methodology. Heliyon. 2019;**5**(11):e02807

[10] Jara AA, Razal RA, Migo VP, Acda MN, Calderon MM, Florece LM, et al. Production of nanocellulose and biocomposites from kawayan kiling (Bambusa vulgaris Schrader ex Wendland) shoots. Ecosystems and Development Journal. 2020;**10**(2):10-17

[11] Okahisa Y, Sakata H. Effects of Growth Stage of Bamboo on the Production of Cellulose Nanofibers. Fibers and Polymers. 2019;**20**(8):1641-1648

[12] Yang K, Li L, Lou Y, Zhu C, Li X, Gao Z. A regulatory network driving shoot lignification in rapidly growing bamboo. Plant Physiology. 2021;**187**(2):900-916

[13] Cesarino I. Unraveling the regulatory network of bamboo lignification. Plant Physiology. 2021;**187**(2):673-675

[14] Li C, Suo J, Xuan L, Ding M, Zhang H, Song L, et al. Bamboo shoot-lignification delay by melatonin during low temperature storage. Postharvest Biology and Technology. 2019;**156**:110933

[15] Zhang Z, Li C, Zhang H, Ying Y, Hu Y, Song L. Comparative Analysis of the Lignification Process of Two Bamboo Shoots Stored at Room Temperature. Plants (Basel). 2020;**9**(10):1399

[16] Jingwen W. Study on Ageing Physiology of Postharvest Bamboo

Shoots. Forestry Science Research. 2002;**15**(6):687-692

[17] Qi X, Ji Z, Lin C, Li S, Liu J, Kan J, et al. Nitric oxide alleviates lignification and softening of water bamboo (Zizania latifolia) shoots during postharvest storage. Food Chemistry. 2020;**332**:127416

[18] Luo Z, Xu X, Yan B. Accumulation of lignin and involvement of enzymes in bamboo shoot during storage. European Food Research and Technology. 2007;**226**(4):635-640

[19] Bugg TD, Ahmad M, Hardiman EM, Rahmanpour R. Pathways for degradation of lignin in bacteria and fungi. Natural Product Reports. 2011;**28**:1883-1896

[20] Zhang J, Li Q, Wei Z. Degradation of bamboo-shoot shell powder by a fungal consortium: Changes in chemical composition and physical structure. International Biodeterioration and Biodegradation. 2017;**116**:205-210

[21] Xiaoli W, Xiaoping G, Mengyun S, Jinjun Y. Study on Contents of Cellulose, Lign in and Activities of POD, PAL in Excised Bamboo Shoots of Phyllostachys edulis. Forestry Research. 2008;**21**:697-701

[22] Yu L, Pei J, Zhao Y, Wang S. Physiological Changes of Bamboo (Fargesia yunnanensis) Shoots During Storage and the Related Cold Storage Mechanisms. Frontiers in Plant Science. 2021;**12**:731977

[23] Li D, Limwachiranon J, Li L, Zhang L, Xu Y, Fu M, et al. Hydrogen peroxide accelerated the lignification process of bamboo shoots by activating the phenylpropanoid pathway and programmed cell death in postharvest storage. Postharvest Biology and Technology. 2019b;**153**:79-86

[24] Shimada N, Munekata N, Tsuyama T, Matsushita Y, Fukushima K, Kijidani Y, et al. Active Transport of Lignin Precursors into Membrane Vesicles from Lignifying Tissues of Bamboo. Plants (Basel). 2021;**10**(11): 2237

[25] Gars ML, Douard NL, Belgacem JB. Cellulose nanocrystals: From classical hydrolysis to the use of deep eutectic solvents. In: Nanosystems. London, UK: IntechOpen; 2019

[26] Klemm D, Heublein B, Fink HP, Bohn A. Cellulose: fascinating biopolymer and sustainable raw material. Angewandte Chemie, International Edition. 2005;**44**:3358-3393

[27] O'sullivan AC. Cellulose: the structure slowly unravels, Cellulose. 1997;**4**:173-207

[28] Liu W, Du H, Zhang M, Liu K, Liu H, Xie H, et al. Bacterial cellulose-based composite scaffolds for biomedical applications: a review. Chemical Engineer. 2020;**8**:7536-7562

[29] Yang Y, Fan F, Xie J, Fang K, Zhang Q, Chen Y, et al. Isolation and characterization of cellulosic fibers from bamboo shoot shell. Polymer Bulletin (Berlin). 2022:1-13

[30] Yang Y, Liu Y, Fang Y, Teng Y, Fang K, Xie J, et al. Extraction and Characterization of Bamboo Shoot Shell Fibers by Sodium Percarbonate Assisted Degumming. Journal of Natural Fibers. 2022:1-10

[31] Yang Y, Zhu M, Fan F, Fang K, Xie J, Deng H, et al. Extraction of bamboo shoot shell fibers by the ultrasound-assisted alkali-oxygen bath method. Cellulose Chemistry and Technology. 2021;**55**:675-680

[32] J. Y. F., S. W.Y., W. L.H., W. H.L., Effects of Ensilage on the preservation

of bamboo shoot shells and their fibre characteristics. Journal of Tropical Forest Science. 2011;**23**:396-403

[33] Zhang W, Fei B, Polle A, Euring D, Tian G, Yue X, et al. Crystal and Thermal Response of Cellulose Isolation from Bamboo by Two Different Chemical Treatments. BioResources. 2019;**14**:3471-3480

[34] Elemike EE, Onwudiwe D, Ivwurie W. Structural and thermal characterization of cellulose and copper oxide modified cellulose obtained from bamboo plant fibre. SN Applied Sciences. 2020;**2**:1-8

[35] Guan M, Zhang Z, Yong C, Du K. Interface compatibility and mechanisms of improved mechanical performance of starch/poly(lactic acid) blend reinforced by bamboo shoot shell fibers. Journal of Applied Polymer Science. 2019;**136**(35):47899

[36] Zhang A, Zou Y, Xi Y, Wang P, Zhang Y, Wu L, et al. Fabrication and characterization of bamboo shoot cellulose/sodium alginate composite aerogels for sustained release of curcumin. International Journal of Biological Macromolecules. 2021;**192**:904-912

[37] Du J, Yang S, Zhu Q, Wu Y, Guo J, Jiang J. Preparation and characterization of thermoplastic starch/bamboo shoot processing by-product microcrystalline cellulose composites. Biomass Conversion and Biorefinery. 2021:1-10

[38] Wu W, Hu J, Gao H, Chen H, Fang X, Mu H, et al. The potential cholesterol-lowering and prebiotic effects of bamboo shoot dietary fibers and their structural characteristics. Food Chemistry. 2020;**332**:127372

[39] Wei Z-J, Thakur K, Zhang J-G, Ren Y-F, Wang H, Zhu D-Y, et al. Physicochemical and Functional Properties of Dietary Fiber from Bamboo Shoots (Phyllostachys Praecox). Emirates Journal of Food and Agriculture. 2017;**29**(7):509-517

[40] Yang K, Yang Z, Wu W, Gao H, Zhou C, Sun P, et al. Physicochemical properties improvement and structural changes of bamboo shoots (Phyllostachys praecox f. Prevernalis) dietary fiber modified by subcritical water and high pressure homogenization: a comparative study. Journal of Food Science and Technology. 2020;**57**:3659-3666

[41] Tang C, Wu L, Zhang F, Kan J, Zheng J. Comparison of different extraction methods on the physicochemical, structural properties, and in vitro hypoglycemic activity of bamboo shoot dietary fibers. Food Chemistry. 2022;**386**:132642

[42] Tang C, Yang J, Zhang F, Kan J, Wu L, Zheng J. Insight into the physicochemical, structural, and in vitro hypoglycemic properties of bamboo shoot dietary fibre: comparison of physical modification methods. International Journal of Food Science and Technology. 2022;**57**:4998-5010

[43] Zhao L, Wu J, Liu Y, Wang H, Cao C. Effect of Lactobacillus rhamnosus GG fermentation on the structural and functional properties of dietary fiber in bamboo shoot and its application in bread. Journal of Food Biochemistry. 2022:e14231

[44] Luo X, Wang Q, Fang D, Zhuang W, Chen C, Jiang W, et al. Modification of insoluble dietary fibers from bamboo shoot shell: Structural characterization and functional properties. International Journal of Biological Macromolecules. 2018;**120**:1461-1467

[45] Luo X, Wang Q, Zheng B, Lin L, Chen B, Zheng Y, et al. Hydration

properties and binding capacities of dietary fibers from bamboo shoot shell and its hypolipidemic effects in mice. Food and Chemical Toxicology. 2017;**109**:1003-1009

[46] Zheng Y, Wang Q, Huang J, Fang D, Zhuang W, Luo X, et al. Hypoglycemic effect of dietary fibers from bamboo shoot shell: An in vitro and in vivo study. Food and Chemical Toxicology. 2019;**127**:120-126

[47] He K, Li Q, Li Y, Li B, Liu S. Water-insoluble dietary fibers from bamboo shoot used as plant food particles for the stabilization of O/W Pickering emulsion Food Chemistry. 2020;**310**:125925

[48] Li X, Guo J, Ji K, Zhang P. Bamboo shoot fiber prevents obesity in mice by modulating the gut microbiota. Scientific Reports. 2016;**6**:32953

[49] Li Q, Fang X, Chen H, Han Y, Liu R, Wu W, et al. Retarding effect of dietary fibers from bamboo shoot (Phyllostachys edulis) in hyperlipidemic rats induced by a high-fat diet. Food & Function. 2021;**12**:4696-4706

[50] Wang N, Wu L, Huang S, Zhang Y, Zhang F, Zheng J. Combination treatment of bamboo shoot dietary fiber and dynamic high-pressure microfluidization on rice starch: Influence on physicochemical, structural, and in vitro digestion properties. Food Chemistry. 2021;**350**:128724

[51] Wang N, Huang S, Zhang Y, Zhang F, Zheng J. Effect of supplementation by bamboo shoot insoluble dietary fiber on physicochemical and structural properties of rice starch: Lwt. 2020;**129**:109509

[52] Okahisa Y, Abe K, Nogi M, Nakagaito AN, Nakatani T, Yano H. Effects of delignification in the production of plant-based cellulose nanofibers for optically transparent nanocomposites. Composites Science and Technology. 2011;**71**:1342-1347

[53] Aumentado HDR, Razal RA, Nacario RC, Manalo MN. Comparative Characterization of Nanocellulose Produced from Kawayan Kiling (Bambusa vulgaris Schrad ex. Wendl) Pulp Using Different Pretreatment Methods. Philippine e-Journal for Applied Research and Development. 2022;**12**:1-11

[54] Rasheed M, Jawaid M, Parveez B, Zuriyati A, Khan A. International Journal of Biological Macromolecules. 2020;**160**:183-191

[55] Lin T, Wang Q, Zheng X, Chang Y, Cao H, Zheng Y. Investigation of the Structural, Thermal, and Physicochemical Properties of Nanocelluloses Extracted From Bamboo Shoot Processing Byproducts. Frontiers in Chemistry. 2022;**10**:922437

[56] Yang Y, Fan F, Fang K, Xie J, Zhang Z, Deng Z, et al. Facile Preparation of All Cellulose Self-reinforced Nanocomposites from Bamboo Shoot Fibers. Research Square. 13 Jul 2021:1-13. DOI: 10.21203/rs.3.rs-449837/v1

[57] Lei Y, Zhang X, Li J, Chen Y, Liang H, Li Y, et al. Nanocellulose from bamboo shoots as perfect Pickering stabilizer: Effect of the emulsification process on the interfacial and emulsifying properties. Food Bioscience. 2022;**46**:101596

Chapter 3

Recent Advances in Bamboo Research, Product Development and Utilisation in Nigeria

Abel Olorunnisola

Abstract

Bamboos are available in all the thirty-six States of the Nigerian Federation, the most common species being *Bambusa vulgaris*. Bamboos are traditionally used for scaffolding in building construction, fencing, staking in farms, handicraft and furniture production, and cooking. Research and development advances in Nigerian universities and research centres in the last twenty years have, however, promoted modern forms of bamboo utilisation in various ways. These include documentation of various properties of locally grown bamboos, development of machines and equipment for bamboo processing, as well as development of modern bamboo products. A few bamboo processing factories have also sprung up in recent years. However, enterprises based on modern bamboo processing techniques are still not common in the country. This Chapter presents a review of the progress made thus far in bamboo processing and product development in Nigeria, discusses the prospects of, and constraints to commercialization of locally developed modern bamboo products, and highlights the way forward.

Keywords: bamboos, bicycle, charcoal, laminated products, roofing tiles, furniture

1. Introduction

Bamboos are some of the fastest-growing plants in the world. Some species grow up to 91 cm within 24 hours, i.e., approximately 4 cm per hour. Different species assume different postures ranging from the erect to the clump forming and the climbing postures. Between the single stemmed and densely clumped forms there are intermediate types with somewhat open clumps [1]. Also, bamboos vary in size from lofty forms with stems up to 23 m high and 23 cm thick, to mere under shrubs [2]. As an engineering material, bamboo has a higher specific compressive strength than wood, brick or concrete and a specific tensile strength that rivals steel [3].

There are over 1000 bamboo species across Asia. However, the five genera commonly found in Africa are *Hiekelia Africana*, *Yushania alpine*, *Oreobambosbuchwaidii*, *Oxytenanthera abyssinica* and *Thannocalamustessellates* [4, 5]. The two main species widely available in Nigeria are *Bambusa vulgaris* and *Oxystenanthera abyssynica*. The former attains a height of between 14 and 20 m at maturity with a girth of about 20 cm, while the later reaches between 8 and 12 m at maturity. The two varieties grow naturally in the forests below River Niger and in Taraba State, mostly around river courses [5]. A 2004 survey [6] showed that Oyo State was one of the most endowed States with bamboo resources in the country, the other States being Ogun, Osun, Ondo, Edo, Delta, Rivers, Akwa Ibom, Cross River, Abia, Ebonyi, Enugu, Anambra, and Imo States. At least 10% of the natural vegetation in these States was dominated by bamboo and that the most common species was *B. vulgaris*.

It has long been recognised that bamboos have multipurpose uses with about 1500 documented uses by 1987 [7, 8]. By 2002, over 2.5 billion people traded in or used bamboo world-wide [9] and documented uses of bamboo had risen to over 10,000 [10]. The ease with which bamboos can be worked, their versatility, strength, and availability recommend them for industrial utilisation which commenced in the 1990s. Some of the modern bamboo products that have since emerged and are on sale in the global market include bamboo plywood also known as plybamboo, oriented strand board, bamboo fibreboard, medium density fibreboard, and laminated bamboo products employed in producing furniture, doors, floor parquets, windows, frames, and partitions [10]. Apart from these products, bamboo is also being used in biomass-based power generation as well as solid bio- fuels, bio-ethanol ad textile production.

However the documented uses of bamboo in Africa are for less than 100, let alone 1000. In general, less than 30 days old culms of some species that are eaten; 6–9 months old culms are used for basketry, mat making, and binding; while 2–3 years old culms are used for handicraft, furniture and construction works [4]. The major uses of bamboo in Nigeria are still limited largely to making scaffolds for building construction (**Figure 1**), fencing, staking in farms, handicraft, and furniture, and as fuel for cooking [5, 7, 8]. Bamboo handicrafts produced across the country include baskets, poultry cages, trays, packaging materials, flower vase, lampshades, sporting goods, gift items and souvenirs et c. Furniture products include a wide variety of living room (**Figure 2**), bedroom, dinning, occasional, infant and garden furniture items. Many of the properties of bamboos that make them suitable for handicraft and art pieces also make them acceptable for furniture making in substitution for wood and rattans which are now becoming scarce due to

Figure 1.
Bamboo culms used for scaffolding.

Figure 2.
Bamboo furniture items locally made in Nigeria.

over-exploitation. Not only are bamboos still available in large quantities in Nigeria, they are much easier to harvest, and transport [11–13]. Their limitations including non-straightness, roughness, and non-uniformity of stalk diameter are often taken advantage of in furniture constructions. The tendency to splitting when bolts, screws and nails, are inserted is easily addressed with the use of special adhesives and fasteners. The relatively high sugar content that makes them susceptible to rot and pests is also readily mitigated with postharvest treatment processes [5, 6, 11].

Whilst industrial utilisation of bamboo has been on the agenda of the Federal Government of Nigeria for some time, little progress has been made towards its actualisation. Industrial enterprises based on bamboo are not yet playing a major role in employment generation in the country [5, 9, 10]. Only a few factories have sprung up in recent years producing toilet rolls and allied products, toothpick, floor parquets etc. with bamboo. Also, lack of appropriate processing machines remains a barrier to the fuller utilisation of bamboo [4, 5, 13].

The objective of this chapter is to review the progress made thus far in bamboo processing and product development in Nigeria, to discuss the prospects of, and constraints to the commercialisation of locally developed modern bamboo products, and to suggest the way forward.

2. Recent advances in bamboo research and product development in Nigeria

Research and development advances in Nigerian universities and research centres in the last twenty years have promoted modern forms of bamboo utilisation. A few examples will be presented here.

2.1 Documentation of basic properties of locally grown bamboo

Quite a number of researchers have documented basic properties of *Bambusa vulgaris* found in different agro-ecological zones in Nigeria. It has been reported that the culm circumference of air-dry 3–5 year old *B. vulgaris* found in the derived Savannah area of the south-west range between 190 and 245 mm, while the culm thickness varies from 6 to 16 mm. The mean dry density of mature (2–5 year old) and over-mature (> 6 years old) *B. vulgaris* found in the Guinea Savannah area of south-west can be as high as 690 kg/m^3 and 620 kg/m^3 respectively. On the contrary, the mean dry density of 5–6 year old *B. vulgaris* found in the moist forest and derived Savannah zones can range from 666.8 kg/m^3 to 745 kg/m^3 [14, 15]. The mean density of 2–4 years old *B. vulgaris* found in the rain forest region of south-west varies between 709.6 and 937.9 kg/m^3 along the culm length from base to top [16]. These findings show that the dry density of mature *B. vulgaris* is about 600 kg/m^3.

The fibre length of mature *B. vulgaris* varies from 2.37 mm to 2.92 mm, and its relatively high cellulose content (61–71%) confirms its suitability for paper production. Its gross calorific value (1810.9–4160.60 cal/kg) also confirms its suitability as a fuel product. The species is moderately resistant to decay fungi and permeable to preservatives and quite strong in shear (6.6–13.1 N/mm^2) [14–16].

2.2 Evaluation of the suitability of local bamboo species for water conveyance

Bamboo is a relatively cheap material for water conveyance compared to other piping materials. Hence, bamboo pipes have used in rural communities for many years for irrigation water conveyance. Studies conducted on hydraulic properties of locally grown *oxytenanthera abyssinica* have confirmed its suitability for use as irrigation pipe [17]. The suitability of *B. vulgaris* pipes for irrigation and drainage has also been confirmed, with a relatively low head loss (0.36% - 0.46%) indicative of low resistance to flow of water, and a friction factor of between 0.037 and 0.049. Affordable locally available joint sealants have also been identified [18–20].

2.3 Development of bamboo culm splitter

Many of the problems associated with bamboo utilisation in the natural form can be alleviated by splitting the culm along the length into individual strips which can then be laminated together to create a number of products. Manual splitting of bamboo is tedious, time consuming and hazardous. Also, the splinted culms tend to be imprecise and irregular in shape. Motorised bamboo splitters available on the international market are not affordable to small-scale bamboo processors in Nigeria. To address this challenge, a manually operated splitter shown in **Figure 3** was developed under the author's supervision at the University of Ibadan, Ibadan, Nigeria [21, 22]. The splitter is user friendly and affordable. A one-off version of the

Figure 3.
Bamboo culm splitter developed at the University of Ibadan, Nigeria.

splitter was produced at cost of US$ 60. It is capable of handling dry and wet culms with and without nodes and can produce six pieces of 300 mm long and 35 mm wide bamboo strips (**Figure 4**) for use in bamboo handicraft and furniture workshops, with a splitting efficiency of 80–85%. As shown in **Table 1**, the average splitting time of the machine is 55 seconds. Culms without nodes are split faster than those with nodes. Also, wet culms are split faster than air-dry culms. The dried and planed bamboo strips (**Figure 5**) were laminated and used in the manufacture of a book-shelf (**Figure 6**).

2.4 Domestication of bamboo bicycle production technology

Bicycles have been in existence for centuries and different materials -iron, alloy steels carbon fibre, titanium and other advanced alloys - have been used in making bicycle frames and fork tubes. With the advent of the Green Movement and their calls for environmental sustainability, bamboos are now being promoted for bicycle production. Different species of bamboo have been used bicycle production in a few African countries including Ghana and Zambia and in many cases sold on the international market. In domesticating the production processes in Nigeria, the first set of bicycles produced with locally available bamboo were made in the African University of Science and Technology, Abuja around 2012 (**Figure 7**). The author also supervised the development of processes for bicycle fabrication with *B. vulgaris* at the Department of Wood Products Engineering, University of Ibadan in 2021 [23]. A sample of the bicycle is shown in **Figure 8**.

Figure 4.
Unplanned bamboo strips produced with the bamboo culm splitter.

S/N	Culm Condition	Culm Thickness (mm)	Nodes Present	Splitting Time (sec)
1	Wet	7.29–12.5	Yes	85–117
2	Wet	7.29–10.42	No	45–65
3	Dry	6.25–13.75	Yes	180–356
4	Dry	6.25–12.5	No	65–120

Table 1.
Effects of culm geometry and condition on splitting time.

Figure 5.
The planed bamboo strips.

Figure 6.
Bookshelf produced with the laminated strips.

Figure 7.
Bamboo bicycles produced at the African University of Science and Technology, Abuja, Nigeria.

2.5 Development of bamboo charcoal and charcoal briquette production facilities

Although bamboos in general have relatively high heat values, volatile contents, lower ash and moisture content, they have traditionally not been valued as firewood

Figure 8.
Bamboo bicycle produced at the University of Ibadan, Nigeria.

Figure 9.
a. the bamboo charcoal kiln with retort open. b. the bamboo charcoal kiln with retort closed. c. the furnace stocked with firewood. d. the bamboo charcoal kiln chimney. e. the bamboo charcoal produced.

since they quickly burn to ashes. Unlike firewood, larger volumes of bamboo are needed for meal preparation. Bamboo charcoal offers a cleaner, more consistent alternative to firewood, in terms of higher energy density and steady combustion [4, 12]. The charcoal industry constitutes a viable source of income in Nigeria because of the relatively low investment cost, simple and adaptable production technology and a ready local market willing to offer competitive prices. Bamboos are easier to process for charcoal making than wood. However, despite the growing concerns about the un-sustainability of wood charcoal production, bamboo charcoal is not yet popular in the country. To facilitate small-scale bamboo charcoal production, a portable kiln was developed and tested by the author (**Figure 9a–e**). The kiln is relatively easy to assemble and simple to use. It consists of two metal barrels and a detachable chimney. The larger barrel serves as the furnace, while the smaller barrel serves as the retort. Evaluation tests have shown that, with the 50 x 60 cm furnace, and 40 x 40 cm retort, 1.5 kg of well-formed lumps of bamboo charcoal can be produced from approximately 6 kg of air-dry *B. vulgaris*. Charring takes about 2 hours, with a cooling time of 2–3 hours. Charcoal production efficiency is about 25%. An experienced operator can operate ten units of the kiln in a day. Its adoption can help to minimise environmental consequences of un-controlled wood charcoal production in the country.

Briquetting is the compaction of particulate materials to highly cohesive fuel products. It is one of the technologies available for converting bamboo into biofuel products. It can also be used to address wastages in bamboo harvesting and processing. For example, branches from clumps, dead poles, the hard and crooked basal parts often discarded during harvesting; trimmings and shavings from furniture makers; and bamboo charcoal fines can all be briquetted. Briquettes are cleaner, easier to package and more convenient to use than firewood and charcoal [12]. They can be produced with manually operated presses or motorised briquetting machines. However, bamboo charcoal briquettes are not yet popular in Nigeria and for small and medium scale operations to succeed in the country, simple, low-cost, manually-operated presses are more appropriate. To address this challenge, the author developed a manual bamboo charcoal briquetting press with throughput capacity of about 50 briquettes per hour (**Figure 10**). The press produces hollow cylindrical fuel briquettes, 10 cm in diameter and 4 cm long, as shown in **Figure 11**). The briquettes are easy to sun-dry and combust in traditional, improved charcoal or purpose-built briquette burning stoves.

If they are conveniently accessible to consumers, bamboo charcoal briquettes could serve as complements to firewood and wood charcoal for domestic cooking and agro-industrial operations. However, for the bamboo charcoal briquetting industry to thrive in Nigeria, the core issues of bamboo cultivation, appropriate product pricing, and awareness creation that would lead to acceptance and wide spread adoption must be addressed. It is likely that consumers might be willing to adopt bamboo charcoal briquettes in replacement for firewood or wood charcoal if the prices are comparable. The natural markets for wood charcoal, i.e., homes, institutional kitchens- schools, prisons, and restaurants- might also be easier to penetrate in promoting the acceptance and adoption of bamboo charcoal briquettes.

2.6 Development of bamboo composite roofing tiles

A roof serves the purpose of protecting a building against sunlight, wind, rain and extreme weather conditions. Different types of roofing materials are available in the global market today, but corrugated zinc roofing sheets remains popular for low-cost building construction in Nigeria. Asbestos-cement roofing sheets have become less popular in the wake of the global campaign against asbestos fibre. Bamboo can be a good substitute for zinc roofing sheets with significant environmental benefits and enormous market potentials. Its suitability for cement-bonded roofing tile production has been acknowledged. The suitability of Nigerian grown *B. vulgaris* fibres for

Figure 10.
The bamboo charcoal briquetting press.

Figure 11.
The bamboo charcoal briquette produced.

the production of durable, relatively strong, and weather-resistant roofing tiles has also been reported [24, 25]. The roofing tiles installed on a building are shown in **Figure 12**.

Another form of bamboo roofing material not yet investigated in Nigeria is the corrugated bamboo roofing sheet typically made from multiple layers of woven bamboo mats impregnated with an adhesive resin (**Figure 13**). The sheets can be produced in a range of sizes to suit particular requirements and can easily be trimmed for special applications. These roofing sheets have the same standard measurements as conventional corrugated roofing sheets and are also potential substitutes to plastic, zinc or corrugated asbestos roofing panels. They are attractive, durable, strong, fire-resistant, and easy to cut and drill. Besides, in comparison with conventional asbestos, zinc and

Figure 12.
Bamboo-cement roofing tiles installation.

Figure 13.
Corrugated bamboo roofing sheet.

plastic roofing sheets (**Table 2**), corrugated bamboo roofing sheets tend to be quieter when it is raining and cooler in hot sun given their relatively high thermal resistance, low thermal transmission coefficient, very good noise insulating properties. Although corrugated bamboo roofing tiles are more water absorbent than asbestos roofing sheets, this challenge can be overcome either with the use of appropriate resins for roofing sheet production or by coating the outer layer of the roofing sheets [26].

In general, the use of bamboo for roofing tiles and sheets not only makes commercial sense, but also has potential long term environmental benefits. It also has

Roofing Sheet Material	Density (g/cm^3)	Water absorption (%)	Thermal resistance m^{2}K/W	Coefficient of thermal transmission W/(m^{2}K)	Noise obstruction compared with the reference board (dB)	Flexural strength (MPa)	Impact strength (MPa)
Bamboo	0.71	36.7	0.030	5.6	9	39.72	9.93
Zinc	1.56	1.7	0.012	6.2	4	N/A*	N/A*
Plastic	N/A*	N/A*	0.007	6.4	0 (benchmark)	N/A*	N/A*
Asbestos	1.60	21.1	0.019	5.9	15	31.91	2.55

**N/A = Not available.*
Source: [26].

Table 2.
Comparison of corrugated bamboo roofing sheet with selected common roofing materials.

Chemical Constituent	Percentage content (%)
CaO	4.43
SiO_2	78.00
Al_2O_3	4.96
MgO	1.02
Fe_2O_3	2.01
K_2O	3.09
MnO_2	0.23
P_2O_5	0.72
TiO_2	0.36

Source: [27].

Table 3.
Chemical constituents of bamboo leaf ash.

potential social impact in terms of boosting income generation people engaged in bamboo cultivation and processing.

2.7 Evaluation of bamboo leaf ash as a pozzolan

Pozzolans are siliceous and aluminous materials that have little or no cementitious properties in themselves but, when finely ground, will chemically react with calcium hydroxide in the presence of water to form compounds that have cementitious properties. In view of global concerns about CO_2 emissions during Portland cement production, efforts have been made over time to identify suitable pozzolans for partial replacement of cement. A well-known pozzolan of organic origin is rice husk ash. However, when properly incinerated, bamboo leaf ash can become pozzolanic. Studies have shown that ashes of Nigerian grown bamboos are pozzolanic since they (i) contain all the main chemical constituents of ordinary Portland cement (**Table 3**). It has been reported that partial replacement of ordinary Portland cement with 10 to 20% of bamboo leaf ash could enhance the workability of concrete [27, 28].

3. Prospects of commercialisation of locally developed modern bamboo products

The prospects of commercialising modern bamboo products in Nigeria are bright for the following reasons:

3.1 Availability of bamboo raw material base

All has not been well with the forestry sector in Nigeria in the last forty years. Nigeria has lost over 60% of its primary forests in the last twenty years and harvesting of at least twenty-six tree species has been banned in different states across the country as far back as 1999 due to their endangered status and concern over their extinction [12]. However, in the midst of the current wood scarcity, bamboos abound. Unlike most timber, bamboo is self-regenerating; new shoots that appear annually ensure future raw material after mature culms are harvested. With a 10–30% annual increase in biomass compared to 2–5% for woody trees, bamboos create a greater yield of raw materials. Besides, bamboo plantations can remain productive for more than 50 years and investments can be fully recovered within 10 years [2–10]. Also, mature locally grown bamboos in the country are large enough to permit industrial utilisation. The locally grown *B. vulgaris* and *oxytenanthera abyssinica* in particular have been found suitable for making sundry industrial products.

3.2 Growing demand for wood products

With population increase, there is a growing unmet demand for wood products in Nigeria partly due to wood scarcity, infrastructural inadequacies and other factors. The challenge of wood scarcity can be readily addressed by substituting wood with bamboo in furniture, floor parquet, wall siding, match stick, tooth pick, plywood, particleboard, fibreboard, tissue paper, writing paper, and paperboard production. Some of these products can be produced from recycled bamboo scaffold [29–32]. Nigeria needs to borrow from the examples of China, India, Malaysia and other Asian countries where planting and conversion of bamboo into fibreboard, particleboard and laminated products to supplement timber in wood products manufacture continue to be encouraged [12].

3.3 Simplicity and adaptability of the production technology

The production technologies for manufacturing numerous bamboo products are similar to those of wood products manufacturing in many respects. Besides, they are suitable for small-to-medium scale manufacturing, hence their special appeal in Nigeria, where Micro, Small and Medium Enterprises (MSMEs) continue to play a big role in the economy and where energy costs are increasing at an alarming rate. Today, there are officially over 41.5 million MSMEs in Nigeria, contributing about 50% of the Gross Domestic Product (GDP) and providing jobs for about 60 million people (i.e., about 86% of the national workforce). It has been well-established that electricity access tends to have lower positive impact on productivity of micro enterprises largely because such firms do not use electricity-dependent machinery and processes. It is also accepted as a general rule that changes in electricity costs do not significantly affect unit costs and sales price in MSMEs. Materials costs, which can account for 80–90% of total production costs, are also often unaffected by electricity outages [33].

3.4 Availability of local expertise

The processes involved in the conversion of the locally available bamboo species into industrial products are well known in the nation's research and development community. There are several Departments of Forestry, Wood Technology, Chemistry, Agricultural, Civil, Materials, Mechanical, and Wood Products Engineering in the country, many of which have experts engaged in bamboo research. The Forestry Research Institute and the Raw Materials Research and Development Council are also at the fore front of bamboo research. There may, therefore, be no need for expatriate services in setting up manufacturing plants.

3.5 Favourable government policies

The success of any industrial enterprise depends largely on a favourable business environment. There is, at present, a proposed national roadmap with the overall objective of optimally tapping the potentials of Nigeria's bamboo resources for job creation, income generation, biomass energy and environmental protection. The extant government policy also encourages enterprises based on locally-sourced raw materials, preferably non-oil based and export-oriented. These policy initiatives constitute a conducive environment for nurturing a viable bamboo processing industry in Nigeria today.

4. Constraints to commercialisation of bamboo product innovations

Research results, inventions and innovations have value only when they serve useful purposes in the society. As at today, the wide array of innovative bamboo processing equipment and products catalogued above are not available in the Nigerian market. The reasons for non-commercialisation of modest breakthroughs in research and development in general in Nigeria are many and varied. They include minimal institutional support for intellectual property activities, and absence of appropriate reward system for developmental research in Nigerian universities [34]. A major constraint to the commercialisation of innovative bamboo products in particular is the unfavourable societal disposition to bamboo as an engineering material. Bamboo has a poor reputation in many local communities as an invasive weed that is difficult to eliminate. Another constraint is the very low level of awareness in the Nigerian public of the local innovations in bamboo processing and utilisation. For example, in a 2020 survey 17% of 40 respondent undergraduate students of the Faculty of Technology, University of Ibadan, Nigeria, aged 20–29 years and 58% of whom are male, had not heard of, or seen a bamboo bicycle before [23]. There is also the constraint of the absence of local bamboo design codes. For bamboo to be acceptable as building components, appropriate engineering design codes must developed. This has not been the case in Nigeria yet, thus hindering its acceptability by Architects and Structural Engineers. Also, there is the dearth of local entrepreneurs, strategic investors and venture capitalists that are willing to invest in the commercialization of bamboo products.

5. The way forward

One way of enhancing industrial utilisation is securing reliable source of raw material. To date, bamboo is still treated as an 'open-access' resource that can be readily harvested from wild forests. More bamboo plantations have to be established and concerted efforts have to be made to promote local investments in bamboo plantations. Harvesting mature bamboo stems is also a crucial part of appropriate bamboo plantation management as it will influence the sustainable development of the plantation. Hence, standards guiding bamboo harvesting have to be developed for effective control over the maturity of items harvested.

It is also important to deepen awareness about recent local developments in bamboo utilisation. The first step towards change is awareness. The second step is acceptance. For bamboo products to be accepted, the general public has to be informed and educated with the intention of influencing their attitudes, behaviours and beliefs towards bamboo as an excellent substitute for wood. For effectiveness, awareness has to be deepened among the following categories of stakeholders:

- Users-people who will use the bamboo products Government at local, state and federal levels who are the policy formulators and regulators

- Influencers-professional associations such as Forestry Association of Nigeria, Nigerian Society of Engineers, etc., non-governmental organisations and people who have the power to influence government policies and decisions

- Providers, including banks and other financial institutions that provide financial resources for investments in bamboo-related projects.

Publication of research briefs, extension notes, etc. by universities and research institutes should be encouraged. Annual technology fairs/exhibitions should also be mounted to show-case new process and product innovations; dialogues with key players in the private sector and cognate government ministries, parastatals and agencies [34]. Current bamboo processors in the country have to be empowered through transfer of adaptable innovative processing technologies that meet international standards. The establishment of the National Bamboo and Rattan Growers, Processors and Marketers Association of Nigeria in October 2018 is a step in the right direction. Finally, for the new structural bamboo products already developed locally to be acceptable as building components, appropriate engineering design codes must be developed. This has to be done to promote their acceptability by Engineers and Architects.

Investment in local power generation is also critical. Insecure electricity supply presently constitutes a serious constraint to the development and expansion of local industries in general. The current electric power generation of around 4 GW in the country is insufficient to meet the country's peak demand of 8.25 GW. It is common knowledge that energy insecurity influences investment decisions since power outages and voltage fluctuation often halt production, damage equipment and affect product quality. This is more important since firm competitiveness usually depends on product quality and the ability to meet orders on time, as well as reduce unit cost of production. Since self-generation is generally more expensive than grid electricity, the best solution is to increase the current generating capacity and to find means of reducing transmission and distribution losses. It may, however, be necessary to focus on reducing technical faults in existing transmission and distribution infrastructure as a short-term priority over the long-term necessity to increase generation capacity.

Greater investments in Vocational and Technical Education (VTE) are also necessary because the country is currently in short supply of well-trained artisans and technicians. Not only are vocational training centres too few, there has also been a decline in enrolments over the years. To promote rapid industrialisation in general and for the bamboo processing industry to take off effectively, the overall disinterest in structured VTE, the inadequacy of basic VTE training centres and infrastructure as well as the dearth of experienced VTE instructors must be addressed.

6. Conclusion

The Nigerian forest estate is no longer capable of supplying the quantity and quality of wood required for industrial production of a myriad of wood products on a sustainable basis. However, there is the possibility of diversification from traditional wood- to bamboo- based products. Bamboo being one of the fastest growing plants makes it a highly attractive natural resource compared to the hardwoods. Several research and development efforts geared towards promoting modern methods of bamboo utilisation in the country have been catalogued. Once the constraints to commercialisation and growth of local industries are removed, the various locally developed modern bamboo products should enjoy good market acceptance in Nigeria as they have done in many other countries across the globe.

Conflict of interest

The author declares no conflict of interest.

Author details

Abel Olorunnisola
Dominion University, Ibadan, Nigeria

*Address all correspondence to: abelolorunnisola@yahoo.com

References

[1] Anokye RR, Bakar ES, Abare AY, Kalong RM, Muhammad A. The difference in density along the bamboo culms of *Gigantochloascortechinii* and *Bambusa vulgaris*. International Journal Emergency Technology Advances in Engineering. 2014;**4**(10):638-643

[2] Liese W. Characterisation and utilisation of bamboo. In: Proceedings of Project Group, P.5-04. Ljubljana, Yugoslavia: IUFRO World Congress; 1986. pp. 11-16

[3] Huang X, Li F, De Hoop CF, Jiang Y, Xie J, Qi J. Analysis of *Bambusa rigida* bamboo culms between internodes and nodes: Anatomical characteristics and physical-mechanical properties. Forest Production Journal. 2018;**68**(2):157-162

[4] Onilude MA. Potentials of bamboo as a raw material for the wood industry in Nigeria. In: A Paper Presented at a Seminar on the Wood and Wood Products Industry Organized by Raw Material Research and Development Council in Conjunction with Manufacturers' Association of Nigeria. Ikeja, Lagos: MAN House; 2005. p. 17

[5] Ogunwusi AA. Unlocking the Potential of Bamboo in Nigeria. Abuja, Nigeria: Venus Multimedia Development Services Ltd.; 2014. p. 203

[6] RMRDC. Bamboo Occurrence and Utilisation in Nigeria. Abuja, Nigeria: Raw Materials Research & Development Council; 2004

[7] Liese W. Research on bamboo. Wood Science and Technology. 1987;**21**:189-209

[8] Atanda J. Environmental impacts of bamboo as a substitute constructional material in Nigeria. Case Studies in Construction Materials. 2015;**3**:33-39. DOI: 10.1016/j.cscm.2015.06.002 2214-5095/$

[9] INBAR. Bamboo in Winter. International Network for Bamboo and Rattan (INBAR) News Magazine. 2002;**8**(2):16

[10] INBAR. South-South in Action-Inspiring Sustainable Development with Bamboo. Beijing: United Nations office for South-South Cooperation & the Network for Bamboo and Rattan (INBAR); 2017. p. 60

[11] Omobowale MO, Ogedengbe K. Trends in fiber characteristics of Nigerian grown bamboo and its effect on its impact and tensile strengths. Bamboo Science Culture. 2008;**21**(1):9-13

[12] Olorunnisola AO. Harnessing the Forester's Harvest for Sustainable Development. Ibadan: University of Ibadan Publishing House; 2013. p. 102

[13] Olorunnisola AO. The potentials of bamboo as a novel building and furniture production material in Nigeria. In: Proceedings of the 5th Nigerian Materials Congress. Abuja: Nigeria Materials Science & Technology Society of Nigeria.; 15-18, November 2006. pp. 180-185

[14] Ogunsanwo OY, Terziev N, Panov D, Daniel G. Bamboo (*Bambusa vulgaris* Schrad.) from moist forest and derived savanna locations in south West Nigeria – Properties and gluability. BioResources. 2015;**10**(2):2823-2835. DOI: 10.15376/biores.10.2.2823-2835

[15] Ogunsile BO, Uwajeh CFU. Evaluation of the pulp and paper potentials of a Nigerian grown *Bambusa vulgaris*. World Applied Sciences Journal. 2009;**6**(4):536-541

[16] Sadiku NA, Oluyege AO, Sadiku IB. Analysis of the calorific and fuel value index of bamboo as a source of renewable biomass feedstock for energy generation in Nigeria. Lignocellulose. 2015;**5**(1):34-49

[17] Umaru BG, Audu I, Ogedengbe K, Omobowale MO. Development of bamboo (*oxytenanthera abyssinica)* stem as irrigation pipe. Journal of Applied Science, Engineering & Technology. 2010;**10**:18-24

[18] Ogedengbe K. Determination of some hydraulic characteristics of bamboo (*Bambusa vulgaris* (Schrad)) pipe. International Science Technology Journal of Namibia. 2014;**3**(1):82-99

[19] Akinbile CO, Fakayode OA, Sanusi KO. Using bamboo (*Bambusa vulgaris*) as a field drainage material in Nigeria. African Journal of Environmental Science and Technology. 2011;**5**(12):1124-1127

[20] Fagbayide SD, Ogedengbe K. Effect of bamboo (*Bambusa vulgaris* (Schrad)) basic joint principles and techniques for irrigation and drainage network water supply system in agricultural systems. International Journal of Agriculture and Food Science. 2017;**6**(17):1037-1045

[21] Abu AT. Design, Construction and Performance Evaluation of a Bamboo Culm Splitter [Thesis]. Nigeria: University of Ibadan; 2020

[22] Oyeleye AE. Modification of a Bamboo Culm Splitter, Design and Construction of a Bookshelf with Bamboo Strips [Thesis]. Nigeria: University of Ibadan; 2021

[23] Adepoju OA, Ayanbiyi OJ. Design and Construction of a Bamboo Bicycle [Thesis]. Nigeria: University of Ibadan; 2020

[24] Olorunnisola AO, Adeniji AO. Investigations on the effects of cement replacement and calcium chloride addition on selected properties of coconut husk fibre-reinforced roofing tiles. Materials Research Proceedings. 2019;**11**:253-259. DOI: 10.21741/9781644900178-21

[25] Omoniyi TE, Akinyemi BA, Olusoji AO. Development of bamboo – Rice husk ash and cement mixture for livestock house roofing sheets. American Journal of Scientific and Industrial Research. 2013;**4**(2):201-209

[26] Corrugated-bamboo-roofing-sheets [Internet]. 2022. Available from: https://www.guaduabamboo.com/blog/corrugated-bamboo-roofing-sheetssite

[27] Umo AA, Odesola I. Characteristics of bamboo leaf ash blended cement paste and mortar. Civil Engineering Dimension. 2015;**17**(1):22-28

[28] Olutoge FA, Oladunmoye OM. Bamboo leaf ash as supplementary cementitious material. American Journal of Engineering Research. 2017;**6**:1-8

[29] Adewole NA, Olayiwola HA. Production of bamboo-lam from flat bamboo strip prepared from Bambusa vulgaris Schrad. In: Proceedings of the 10th Nigerian Materials Congresses. Akure, Nigeria: Materials Science & Technology Society of Nigeria; 21-24 November 2011. pp. 346-350

[30] Adewole NA, Bello KO. Recycling of bamboo (*Bambusa vulgaris* Schrad) recovered from scaffold into material for furniture production. Innovative Systems Design and Engineering. 2013;**4**(9):73-79

[31] Adewole NA, Etta PN, Ibor BB. Characteristics of selected properties of parquet laminate produced from *Bambusa vulgaris* Schrad harvested

from Akamkpa local government area, Calabar, Nigeria. International Journal of Emerging Trends in Engineering and Development. 2017;**3**(7):183-192

[32] Abimbola MB. The Prospect of Using Bamboo as an Alternative to Wood for Match Stick Production in Nigeria. A seminar report,. Nigeria: University of Ibadan; 2020

[33] Olorunnisola AO. Nigerian Energy Situation and its Impact on MSME. A Keynote Address at a 5-Day Capacity Building Workshop by the National Centre for Energy Efficiency and Conservation. Nigeria: University of Ibadan; 14 February, 2022. p. 5

[34] Olorunnisola AO. Challenges of commercialising engineering research outputs from Nigerian universities. Acta Academica Supplementum. 2011;**2**:217-238

Chapter 4

The Physical and Mechanical Properties of Compreg Laminated Bamboo Strips Lumber (LBSL) of *G. scortechinii*

NurIzzaati Saharudin, Norhafizah Saari, Othman Sulaiman and Rokiah Hashim

Abstract

The chapter explores the effect of resin concentration and pre-drying time on the physical and mechanical properties of compreg-laminated bamboo strips lumber (LBSL) from G. scortechinii species. The compreg LBSL panels were manufactured using two concentrations of low-molecular-weight phenol formaldehyde (LMwPF) resin, that is, 100 and 70% at three different pre-drying times (12, 18, and 24 h). Based on the findings, the panel with 70% of LMwPF required a longer time and higher temperature to cure than the panel with 100% LMwPF. The weight percent gain (WPG) and moisture content of the panels increased in line with increasing resin concentration but decreased with increasing pre-drying time. The opposite trend is indicated by density. As for the mechanical properties, the strength of the panel increases along with the increase in resin concentration and pre-drying time. However, for high-concentration resins, prolonging the pre-drying time reduces the strength of the sample. The highest flexural strength and tensile strength were observed in the 100/18 sample with values of 260 and 27 MPa, respectively. The results for formaldehyde emissions show panels with lower resin concentrations: 70% LMwPF and prolonged pre-drying time reduce formaldehyde emission levels.

Keywords: compreg, bamboo, mechanical, formaldehyde emission, physical

1. Introduction

Bamboo is limited and less practiced due to heterogeneity in shapes, properties, and lack of standards [1]. Its hollow tubes with high silica content on the surface are difficult to join and adhered [2]. The thin and uneven culm wall thickness with defects caused by fungal and insects attacks worsen the condition. The emergence of new technologies has sparked the transformation of bamboo to engineered bamboo for further extended uses. A combination of bamboo in the form of veneer strands or strips with an adhesive undergo hot pressing or densification could overcome the shortcomings inherited by raw bamboo and possibly provide high performance for its end use [1, 2]. Despite having mechanical properties that are comparable to modern structure materials, their properties are greatly influenced by the species and manufacturing method used [3].

A lot of studies have been conducted to determine the appropriate methods for producing high-performance engineered bamboo. Compreg is one of the effective technique to improve lignocellulosic materials by an impregnated synthetic resin, such as low molecular weight phenol formaldehyde (LMwPF) resin, followed by hot-pressing densification [4]. The blockage of the hydroxyl groups with resin, and increment of the panel density parallel with its compressibility had enhanced their dimensional stability, durability, and strength properties of the panel produced [5, 6]. Studies show that compreg can increase up to two- to three-fold of wood's original strength due to the increase in density [4–7]. However, a major concern of this compreg wood is the emission of formaldehyde.

Formaldehyde emission (FE) is free formaldehyde released by methylol groups in the oligomeric chain when the resin is incompletely cured [7]. It has a negative impact on human health. Previous studies had shown that LMwPF resin released significantly higher free formaldehyde than commercial PF resin [8]. Various methods are taken to derive the FE value of the final panel that was made up of low-concentration resins [9]. However, the presence of high-water molecules in diluted LMwPF resin had decreased its solid content and thus lowered its FE [8].

Consisting of small molecules and short-chained, LMwPF resin requires a longer time to cure than commercial PF. In the case of LMwPF, heat is applied to remove water molecules to form a polymer crosslink in order to cure the resin. However, LMwPF resin takes a longer time to cure compared with other resins because it needs to be transformed to oligomer before end-up with complete polymer linkage [10]. The addition of water molecules in the resin solution increases its curing time. The curing process is vital to produce high-quality panels. A proper rate of time is needed to create a high bonding strength between the substrates. This is because panel failure can occur prematurely when the stress value at that location goes beyond the glue bonding strength sooner than its bottom fiber reaches its tensile strength [11].

Regarding these facts, the objective of this study was to investigate the effect of resin concentration and pre-drying time on the physical and mechanical properties as well as formaldehyde emission of compreg LBSL of *G. scortechinii.*

2. Materials and methods

2.1 Materials preparation

The matured culms of *G. scortechinii,* aged 3 years old with a height of 2.4 m were purchased from the bamboo biocomposite (M) Sdn. Bhd. in Gerik, Perak. The culms are divided into three parts of equal length, namely bottom, middle, and top and cross-cut. Each portion was split into eight parts and peeled off the outer and inner layers using a planner into a strip with a minimum thickness of 0.5 cm. Thereafter, the strips were dried in a kiln dryer to a relative MC of 5–10% at an ideal temperature set at 70°C.

2.2 Resin formulations

The low molecular weight of phenol formaldehyde (LMwPF) resin was obtained from Malaysian adhesive chemicals (M) Sdn. Bhd. in Shah Alam, Selangor. The resin was divided into two types, Type I and Type II. Type I resin was an original formulation resin, while Type II resin was a modified resin of Type I. Type II was prepared by diluting Type I resin with distilled water with a ratio of 30:70 based on Hoong et al. [8]. The solid content and resin curing properties were determined and carried out based on the previous studies [12, 13].

2.3 Manufacturing process

The manufacturing process of compreg laminated bamboo strips lumber (LBSL) of *G. scortechinii* was modified from previous studies [14, 15]. The dried strips of *G. scortechinii* were cut into 25 cm in length and vacuumed for 2 h before being immersed overnight in the resin solution (Type I and Type II). Thereafter, the impregnated strips were removed and pre-dried in the oven at 60°C for three different duration times (12, 18, and 24 h). The strips were arranged in parallel and hot-pressing up to half of their original thickness for 20 min at 150°C and 20 MPa and followed by cold-pressing. The conditions of compreg LBSL panel manufacturing process were presented in **Table 1**. The panel was conditioned in a conditioning room at 25°C and 65% (±2%) prior to properties evaluation.

2.4 Evaluation of physical properties of compreg LBSL of *G. scortechinii*

The density and moisture content (MC) of the sample with dimension of 5×5×1 cm was determined according to BS EN 323 and BS EN 322 based on Eq. (1) [16, 17].

$$\text{Density} = \text{m}/\text{v} \tag{1}$$

where m is mass of the sample and v is the volume of the sample, while the MC of the sample is calculated using Eq. (2):

$$\text{MC,\%} = (\text{Wa–Wb}/\text{Wb})\ \text{x}\ 100 \tag{2}$$

where Wa represents the initial weight of the sample and Wb represents the oven-dried weight of the sample.

Sample	Type of resin	Pre-drying time (h)
100/ 12	Type I	12
100/ 18	Type I	18
100/ 14	Type I	24
70/ 12	Type II	12
70/ 18	Type II	18
70/ 24	Type II	24

Table 1.
Manufacturing conditions of compreg LBSL panel made from G. Scortechinii.

2.5 Evaluation of mechanical properties of compreg LBSL of *G. scortechinii*

The modulus of rupture (MOR), modulus of elasticity (MOE), and compression parallel to the grain of the panel were determined according to ASTM 5456 with slight modifications on the sample size [18]. The sample was prepared in size of 20.0×2.5 cm for bending and 10×2.5 cm for compression parallel to the grain and tested by employing Instron machine model UTM-5582. The tensile shear strength was carried out based on EN 302-1-2004 [19]. A total of six samples per condition were prepared for each test.

2.6 Analysis of formaldehyde emissions of compreg LBSL of *G. scortechinii*

The formaldehyde emission test was carried out using the desiccator's method as specified in the Malaysia Standards (MS 2005) of wood-based panels [20]. Samples with a total area of approximately 1800 cm^2 were hung in a container containing 300 ml of distilled water at a height of 4 cm from the water level for 24 h at room temperature. The formaldehyde absorbance was measured photometrically at 412 nm wavelength after 24 h. The formaldehyde emission was determined by Eq. (3) as below:

$$G = f\ (Ad\text{–}Ab)\ X\ 1800/S \tag{3}$$

where G is a concentration of formaldehyde due to the test piece (mg/L), Ad is the absorbance of the solution from the desiccator containing test sample, Ab is the absorbance of the background formaldehyde solution, f is the slope of calibration curve for the standard formaldehyde solution, (ppm), and S is the surface area of the test pieces (cm^2).

3. Results and discussions

3.1 Curing characteristics of LMwPF resin

Based on **Figure 1**, both types of resins showed a single exothermic peak, which refers to complete thermal curing without post-curing resulting from homopolymerization [21]. Both resins undergo thermal healing reactions at 28°C with the highest exothermic peaks for Type I and Type II occurring at 89.46 and 93.29°C,

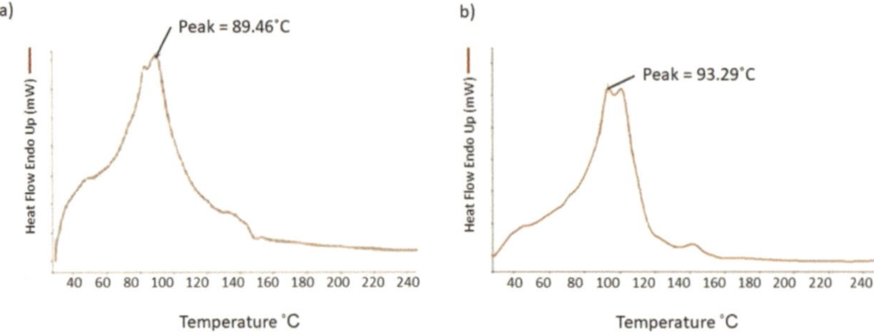

Figure 1.
The DSC analyses of LMwPF resin at different formulations (a) Type I resin and (b) Type II resin.

respectively. The highest exothermic peak indicated the beginning of the curing process where the water molecules are released. Then, the resin undergoes gelation due to the rigid crosslink network and subsequently turns to glass and decomposed as the temperature increases [12, 13]. The curing time for Type I and Type II resin were approximately 15 and 16 min (10°C/min) as both undergo thermal curing reactions up to 148 and 158°C, respectively. Type II resin required higher temperatures and longer time for polymer condensation to form a cross-linked network than Type I due to its less solid content, which is 32%. Moreover, Type II resin contained more water, which acted as an energy barrier for curing process [10, 22].

3.2 Weight percent gain (WPG) of compreg LBSL of *G. scortechinii*

The mean WPG of the panel is shown in **Table 2**. The highest WPG of the panel was detected by sample 100/12, while the lowest was by sample 70/24, with mean values of 26.14 and 15.08%, respectively. The higher solid content of the resin had increased the weight of the sample made from Type I resin. The highly concentrated resin is capable of retaining more polymers in the treated sample compared to low-concentration resin. However, this low-concentration resin known as thin resin, can simply penetrate the cell wall and pass the entire layers of wood cells easily but is difficult to retain in it due to its lower viscosity [9, 23]. Besides that, the pre-drying time also affects the weight of the sample. Based on the results, the mean WPG of compreg sample was inversely proportional to the pre-drying time. Both types of resins showed similar trends. This might be related to the properties of the resin. As a waterborne resin, LMwPF resin releases water during the curing process and prolonged pre-drying time. Thus, more water to be removed and more cross-linked polymers will form. This caused reduction in sample weight [7, 24].

Sample	WPG (%)	MC (%)	Density (g/cm^3)
100/12	26.14	8.67	0.924
	(−4.57)	(−0.14)	(−0.004)
100/18	24.65	8.62	0.876
	(−4.59)	(−0.14)	(−0.008)
100/24	18.91	8.17	0.874
	(−2.75)	(−0.15)	(−0.004)
70/12	16.84	8.82	0.789
	(−2.24)	(−0.23)	(−0.004)
70/18	16.57	8.65	0.799
	(−2.95)	(−0.13)	(−0.005)
70/24	15.08	8.29	0.734
	(−4.56)	(−0.22)	(−0.009)

Note: Values in parentheses are standard deviation.

Table 2.
The physical properties of the compreg LBSL of G. Scortechinii.

Sample	MOR (MPa)	MOE (MPa)	Compression (MPa)	Tensile (MPa)
100/12	260.18 ± 3.22	24097.53 ± 375.50	129.57 ± 1.40	25.62 ± 0.48
100/18	268.33 ± 10.16	25805.50 ± 805.67	130.87 ± 1.41	26.64 ± 0.56
100/24	257.55 ± 10.64	28397.23 ± 676.16	134.18 ± 1.69	25.58 ± 0.57
70/12	217.55 ± 2.35	21361.73 ± 596.00	120.77 ± 1.02	23.82 ± 1.11
70/18	235.70 ± 1.86	23611.95 ± 1319.06	125.65 ± 1.69	24.21 ± 0.58
70/24	237.63 ± 5.36	24299.65 ± 926.32	126.87 ± 2.12	24.27 ± 1.02

Data expressed as an average ± standard deviation.

Table 3.
The mechanical properties of compreg LBSL of G. Scortechinii.

3.3 Moisture content (MC) and density of compreg LBSL of *G. scortechinii*

The MC of the panel was inversely proportional to the resin concentration and pre-drying time, but the density was the opposite. The density of the panel increased with the increase in resin concentration but decreased with the increase in pre-drying time. Based on **Table 3**, the sample of 70/12 had highest MC value, 8.82%. Meanwhile, the highest value of density was detected in a sample of 100/12, 0.924 g/cm^3. According to Purba et al., density and WPG are interrelated in which the increase in WPG will increase the sample density since the resin will soften the cell wall and thus facilitates its compaction [9]. It is contradicted with samples made from low-concentration resins, where the resin tends to outflow the sample during hot-pressing densification, as it is highly saturated with water, resulting in reduced density [25]. However, the MC of the panel made from LMwPF showed an increment due to its high-water content. Meanwhile, prolonging the pre-drying time decreased the MC and density of the sample because more water evaporated and accelerated the curing process, which minimized the compression rate [7, 9].

3.4 Mechanical properties of compreg LBSL of *G. scortechinii*

Mechanical test results were presented in **Table 3**. Based on the results, the highest MOR was observed in sample 100/18 with a value of 260 MPa, while the lowest MOR was observed in sample 70/12 with a value of 218 MPa. Sample 100/24 had highest elasticity (MOE) followed by sample 100/18 with values of 28,397 and 25,806 MPa, respectively. The MOR and MOE of the samples increased in line with the increase in resin concentration and pre-drying time. However, prolonging the drying time, reduced the MOR of the sample, especially for highly concentrated resins. This is because the presence of the resin in the strips had caused the transformation of the strips into a plastic-like material [26]. The retention of resin in bamboo strips as a result of cross-linked formation between it and the substrate during the pre-drying process and hot-pressing densification as well enhanced the MOR and MOE of the sample [8, 27]. However, prolonged pre-drying time causes the resin to excessively cure and become fragile. This was due to high crosslinking in the strips themselves, which weakens the bonding between the layers, and consequently weakens the sample due to poor glue line [5].

The compression parallel to the grain of the sample showed a similar trend with bending strength, which increased with increasing resin concentration and pre-drying time. The samples with high resin content in the lumen and cell walls had higher compressive strength [28]. Higher polymer loading will form a stronger cross-linked thermoset polymer, which is strong, insoluble, and hard to melt and soften [9, 28]. Sample with resin concentration of 100% (Type 1) and pre-drying time of 24 h showed the highest values of compression, 134 MPa followed by sample 100/18 with a value of 131 MPa.

The tensile strength showed the highest reading for sample 100/18 with the value of 27 MPa due to its higher resin concentration. Meanwhile, the lowest was recorded in sample 70/12 with a value of 24 MPa. The lower concentration resins contain high water molecules and these result in starved glue line. Thus, causes a weak bonding between substrates due to the lack of resin retained on the surface other than the limited linkage formation between the substrate [8]. Since the resin compreg relies heavily on the resin emitted by the impregnated strips during hot-pressing densification, it is important to ensure that is sufficient for strong bonding. The pre-drying time increases the tensile strength of the sample but lengthening the pre-drying time will weaken the sample bonding since the resin polymerization process is fully formed in the lumen and this minimizes the formation of cross-linking polymer between the strips [7].

3.5 Formaldehyde emission of compreg LBSL of *G. scortechinii*

The formaldehyde emission (FE) of the panel is shown in **Figure 2**. The highest FE was observed on sample 100/12 with a value of 15.62 mg/L. The sample was made from Type I resin and had undergone 12 h pre-drying times. The highest value of FE might be due to the high solid content of the resin in the sample [7–9]. During polymerization process, formaldehyde formed a linkage with the substrate. The incomplete curing leads to the increase of FE since free formaldehyde released from methylol groups had increased [8]. Prolongs pre-drying time will increase the resin polymerization, especially in a low molecular weight resin, which contains more short-chain oligomers in the system, resulting in decreased FE from the sample [26].

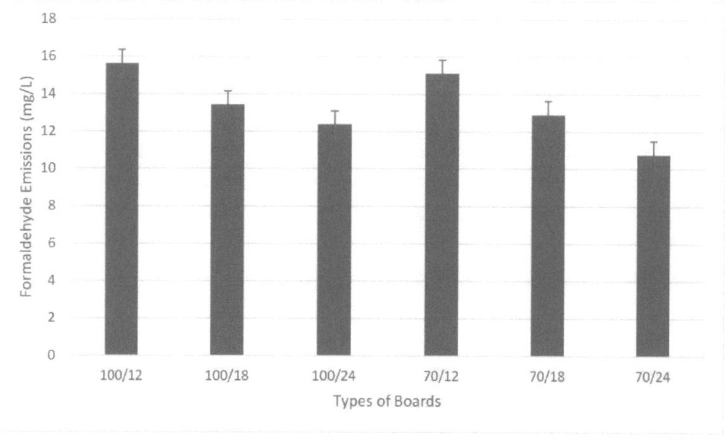

Figure 2.
Formaldehyde emission of compreg LBSL of G. scortechinii *sample.*

4. Conclusions

The resin concentration and pre-drying time have a significant effect on the strength, density, and formaldehyde emission of compreg LBSL of *G. scortechinii* sample. The increase of resin concentration, increased the weight and density of the panel. Thus, increased strength as well as density is directly proportional to the strength of the panel. The pre-drying time showed a positive effect on the properties of compreg LBSL of *G. scortechinii* panel. Type I resin with a high concentration of resin and 18 h of pre-drying time gave optimum results. However, in terms of formaldehyde released, Type II resin with pre-drying time of 24 h showed some reduction of FE from compreg LBSL of *G. scortechinii* panel.

Acknowledgements

The authors would like to acknowledge Universiti Sains Malaysia for providing a research university grant (1001/PTEKIND/846108) and Ministry of Higher Education, Malaysia for the SLAB scholarship throughout NurIzzaati Saharudin studies.

Author details

NurIzzaati Saharudin*, Norhafizah Saari, Othman Sulaiman and Rokiah Hashim
Division of Bioresource, Paper and Coatings Technology, School of Industrial Technology, Universiti Sains Malaysia, Penang, Malaysia

*Address all correspondence to: nurizzaati@usm.my

References

[1] Sharma B, Gatóo A, Ramage M. Effect of processing methods on the mechanical properties of engineered bamboo. Construction and Building Materials. 2015;**83**:95-101

[2] Xing WJ, Hao I, Galobardes S, Wei Z, Chen KS. Engineered bamboo's further application: An empirical study in China. In Proc., MATEC Web of Conf. Paris: EDP Sciences. 2018

[3] Mahdavi M, Clouston PL, Asce AM, Arwade SR. Development of laminated bamboo lumber: Review of processing, performance and economical considerations. Journal of Materials in Civil Engineering. 2011;**23**:1036-1042

[4] Zaidon A, Bakar ES, Paridah MT. Compreg laminates from low density tropical hardwoods. Proceedings of the International Convention of Society of Wood Science and Technology and United Nations Economic Commission for Europe—Timber Committee. 11–14 October (2010), Geneva

[5] Anwar UMK, Paridah MT, Hamdan H, Bakar ES, Sapuan SM. Impregnation and drying process of treated bamboo strips with low molecular weight phenol formaldehyde (LMwPF) resin. Journal of Polymer Materials. 2008;**25**:265-273

[6] Shams MI, Yano H. Development of selectively densified surface laminated wood-based composites. European Journal of Wood, and Wood Products. 2009;**67**:169-172

[7] NurIzreen FA, Zaidon A, Rabia'tolAdawiyah MA, Bakar ES, Paridah MT, Hamami SM, et al. Enhancing the properties of low-density hardwood *Dyeracostulata* through impregnation with phenolic resin admixed with formaldehyde scavenger. Journal of Applied Science. 2011;**11**: 3474-3481

[8] Hoong YB, Loh YF, Chuah LA, Juliwar I, Pizzi A, Paridah MT, et al. Development a new method for pilot scale production of high-grade oil palm plywood: Effect of resin content on the mechanical properties, bonding quality and formaldehyde emission of palm plywood. Material Design. 2013;**52**: 828-834

[9] Purba TP, Zaidon A, Bakar ES, Paridah MT. Effects of processing factors and polymer retention on the performance of phenolic-treated wood. Journal of Tropical Forest Science. 2014; **26**:320-330

[10] Haupt RA, Sellers TR. Characterizations of phenol-formaldehyde resol resins. Industrial Engineering Chemistry Research. 1994; **33**:693-697

[11] Wan Tarmeze WA. Numerical analysis of laminated bamboo strip lumber (LBSL), [PhD thesis], School of Engineering, University of Birmingham, United Kingdom. 2005. p. 239

[12] Nor Hafizah AW, Paridah MT, Yuziah NMY, Zaidon A, Adrian CCY, Nor Azowa I. Influence of resin molecular weight on curing and thermal degradation of plywood made from phenolic prepreg palm veneers. Journal of Adhesion. 2014;**90**:210-229

[13] Anwar UMK. Impregnation of bamboo (*Gigantochloa scortechinii*) with phenolic resin for the production of dimensionally stable plywood. [PhD thesis]. Universiti Putra Malaysia, Serdang, Selangor. 2008

[14] Bakar ES, Jun H, Zaidon A, Adrian CCY. Durability of phenolic-resin-treated oil palm wood against subterranean termites and white-rot fungus. International Bio-deterioration and Biodegradation. 2013;**85**:126-130

[15] Chong YW, Bakar ES, Zaidon A, Hamami SM. Treatment of oil palm wood with low-molecular weight phenol formaldehyde resin and its planing characteristics. Wood Research Journal. 2010;**1**:7-12

[16] BS EN 322. Wood-based panels. Determination of moisture content. 1993

[17] BS EN 323. Wood-based panels. Determination of density. 1993

[18] ASTM D5456. Standard specification for evaluation of structural composite lumber products. Annual Book of Standards. Vol. 4(10). West Conshohocken, PA: ASTM International; 1999

[19] European Committee for Standardization. EN 302–1 2004, Adhesives for Load-bearing Timber Structure- Test Methods- Part 1- Determination of Resistance to Delamination. Brussels: CEN; 2004b

[20] Malaysian Standard. MS 1787: Part 15. Determination of formaldehyde emission by desiccator method. Putrajaya: Department of Standards Malaysia; 2005

[21] Liu XQ, Huang W, Jiang YH, Zhu J, Zhang CZ. Preparation of a bio-based epoxy with comparable properties to those of petroleum-based counterparts. Express Polymer Letters. 2012;**6**:293-298

[22] Xing C, Zhang SY, Deng J, Wang S. Urea formaldehyde resin gel time as affected by the pH value, solid content and catalyst. Journal of Applied Polymer Science. 2007;**103**:1566-1569

[23] Guan M, Yong C, Wang L. Microscopic characterization of modified phenol-formaldehyde resin penetration of bamboo surfaces and its effect on some properties of two-ply bamboo bonding interface. BioResources. 2014;**9**:1953-1963

[24] Roziela Hanim A, Zaidon A, Anwar UMK, Rafidah S. Effects of chemical treatments on durability properties of *Gigantochloa scortechinii* strips and ply-bamboo. Asian Journal of Biological Sciences. 2013;**6**:153-160

[25] Khairunnisha IPN, Bakar ES, Azwa AN, Choo ACY. Effect of combination oven and microwave heating in the resin semicuring process on the physical properties of 'Compreg' OPW. BioResources. 2014;**9**: 4899-4907

[26] Amarullah M, Bakar ES, Zaidon A, Mohd HS, Febrianto F. Reduction of formaldehyde emission from phenol formaldehyde treated oil palm wood through improvement of resin curing state. Journal of Tropical Wood Science and Technology. 2010;**8**:9-14

[27] Kultikova EV. Structure and properties relationships of densified wood, [thesis]. Virgina: Virginia Tech, Blacksburg. 1999; p. 136

[28] Zaidon A, Kim GH, Bakar ES, Rasmina H. Response surface methodology models of processing parameters for high performance phenolic compreg wood. Sains Malaysiana. 2014;**43**:775-782

Chapter 5

Advances in Bamboo Composites for Structural Applications: A Review

Medha Mili, Anju Singhwane, Vaishnavi Hada, Ajay Naik, Prasanth Nair, Avanish Kumar Srivastava and Sarika Verma

Abstract

The fastest-growing plant on earth is bamboo; it grows three times as quickly as most other species and is a renewable, adaptable resource with high strength and lightweight. Bamboos are a valuable alternative resource with high physical similarities with genuine hardwoods. Using these naturally available renewable bamboo resources provides a practical approach to an eco-friendly industry mainly based on green materials and sustainable technologies with minimum impact on nature. In this regard, developing advanced bamboo-based composites is an attractive step. The extensive use of bamboo composites is a result of their advantageous qualities, including dimensional stability, natural colour, exquisite texture, and ease of manufacturing. The bamboo-based composites have immense potential to perform as a wood substitute that can reduce timber import and meet future timber requirements that are presently fulfilled by cutting trees or importing timber. This chapter aims to exhibit these advanced bamboo composites as a competitive and sustainable substitution for conventional timber material for structural applications. The present chapter highlights the advanced bamboo composites as engineered materials utilised mainly for structural applications in housing sectors and construction industries in the form of standard regular shapes such as beams, planks, lumbers, truss elements etc. One of the sections would be dedicated to the future scope of these advanced bamboo composites and recommendations.

Keywords: bamboo, renewable resources, structural applications, housing sectors, construction industry

1. Introduction

Bamboo is giant woody herbage traditionally used for developing fences, furniture and construction, and financial and social roles. Green Gold and Poor Man's Timber are some of its other names. Due to several serious issues and the restrictions on the felling of trees, there is a growing interest in sustainable materials for structural applications worldwide. It is believed that bamboo is a durable and highly renewable resource that may be farmed in any region with a temperate environment [1].

However, the selection of structural materials can considerably impact the environment. Global worry over the greenhouse effect is growing as the earth's surface temperature rises. The excessive release of harmful chemicals and carbon dioxide into the atmosphere is to blame. Bamboo growth should also be taken into consideration as a solution for an accurate interpretation of the carbon balance in an ecosystem since carbon dioxide is the main cause of this heightened greenhouse impact [2, 3]. In this regard, bamboo and wood are choices as both can absorb carbon dioxide as they grow. Wood can have a high sustainability rate because it is inherently renewable, recyclable, and reused [4–6]. In addition, carbon may be stored for a long time in bamboo and wood constructions. The fastest-growing or herbaceous biogenic materials—also known as fast-growing or herbaceous biomass—are among the various biogenic materials and hold the greatest promise for reducing climate change because they can store carbon much more quickly than trees, whose high carbon content is typical of them, [7–9]. Bamboo, as a sustainable material, meets the requirements of green architecture. As for architects using bamboo was not just technical but also a deep consideration of traditional culture, ecological consciousness, thermal performance, emotional dimension, commercial and many other factors. As is well known, South China is home to two important commercial tree species: Chinese fir (*Cunninghamia lanceolata* (Lamb.) Hook.) and Moso bamboo (*Phyllostachy pubescens*). In China, structural timber use and research have received more attention over the last two decades. Due to the high quality of its timber and their economic importance [10–12]. Consequently, the bamboo species have been found to have a lot of potential for the building sectors in the future because their derived products have more sustainability rates and are naturally renewable, recyclable, and reusable, [13]. One of these is forest products, and the increased demand for them will necessitate their expansion and exploitation with little waste and loss. Growing environmental consciousness and dwindling natural resource availability have sparked a need for environmentally friendly materials in an effort to keep the price of conventional synthetic fibre reinforced composites down. Due to its low cost and low energy consumption, natural fibre reinforced green composites - originally developed for the automobile industry-has recently attracted significant scientific interest [14].

Bamboo possesses superior physical and mechanical characteristics similar to hardwoods and is thus considered a potential substitute for wood [15]. Bamboo is a good material for construction in columns and beams, walled envelopes, shear walls, ground materials, and other purposes because of its extraordinary strength-to-weight ratios and optimum thermal, acoustic, and other qualities with high tensile strength [16]. It has a much longer history of use as a construction material than wood, making it an essential component of daily life [17]. Bamboo's quick rate of growth is one of its advantages. By growing between 20 and 100 cm during a day, it can reach its full height of 15–30 m in two to four months [18]. Unlike wood, which takes over 20 years to reach maturity, it does so in just 3–4 years. When fully grown, bamboo has tensile strength that rivals mild steel. It can grow in currently unproductive regions (like on an eroding hillside), and because its root system keeps growing even after harvest, it can produce new shoots. Except for alkaline soils, deserts, and marshes, bamboo may grow on plains, hilly terrain, and high-altitude mountainous areas. According to some research, bamboo yields more biomass than other lignocellulosic crops, growing at a rate of 15-30 m height in couple of months [7, 9, 19]. Therefore, bamboo is more cost-efficient and effectively reduces the greenhouse effect when used as a structural material. In contrast, bamboo is an anisotropic material forming a seismic perspective and an excellent option for structures like bridges erected in seismically active areas due to its lighter weight, viscoelastic qualities, and deformability than steel or concrete. The knots, spiral grains, and wood twists are a few significant elements that affect bamboo's mechanical properties, composed of cellulose, hemicellulose, lignin, and extractives [20, 21]. India's bamboo species mainly suitable for composites are: *B. bamboos, B. Tulda, B. vulgaris, B. balcoa, B. mutans, B. polymorpha, Dendrocalamus strictus, D. asper, M. bacciffera.* However, the tiny diameter of bamboo culms and the wide range of its mechanical qualities limit the use of natural bamboo as a building material. Composite materials made of bamboo have been created to get around these limitations [22].

This chapter will introduce advanced bamboo composites, engineered materials primarily used for structural applications in the housing and construction sectors. These composite materials are typically used in standard regular shapes like beams, planks, lumber, and other components. The primary goal of this chapter is to impart the fundamental knowledge necessary for developing bamboo composites and their use in structural applications. Additionally, it emphasises the critical notion of methods and technology for bamboo composites and their numerous structural applications. One section contains the primary details regarding bamboo and engineered bamboo composites. Another section focuses on improving the understanding of different bamboo composites used for structural purposes. The conclusion and recommendations for further improvement in this area of research are also highlighted.

2. Bamboo - its composition and characteristics

Like wood, bamboo is a naturally occurring organic substance that is heterogeneous and anisotropic. However, there are notable variances in their morphology, structure, and chemical makeup, highlighting particular physic mechanical capabilities. Bamboo has more strength, tremendous toughness, and high stiffness than wood. Bamboo is a very robust material. Its tensile and compressive strengths are greater than those of the majority of other woods. Bamboo can sustain severe stress and has an excellent strength to weight ratio and structural integrity. Bamboo is a very strong material that is used to make flooring, cabinets, furniture, and buildings. Bamboo has the benefit of being "simple to produce." Although bamboo plants are not the easiest plants to propagate, they are relatively simple to plant and flourish in the landscape. It would be simpler if we could just gather the seeds and grow them from there. Bamboo is a popular material because of this. Generally speaking, they have robust features, yet they could be considered weak from a different perspective. Therefore, it is essential to comprehend bamboo resources for their effective utilisation. **Figure 1** shows the unique properties of bamboos.

2.1 Different bamboo structural forms as a building material

When designing with bamboo, whether for furniture, sculpture, or architecture, there is a constant urge to use the material's organic, flowing shapes. However, it's not like all bamboo poles develop as bent poles; some wobble and even twist as they grow, while the majority are rather straight. However, bamboo is frequently used in curving accessories, furniture, construction materials like windows and doors, and integrated building structures. **Figure 2** illustrates how bamboos are handled such that as they grow, they adopt the proper shapes and structures: This is accomplished by moulding into shape using a variety of techniques, such as tying, bolting, and glueing,

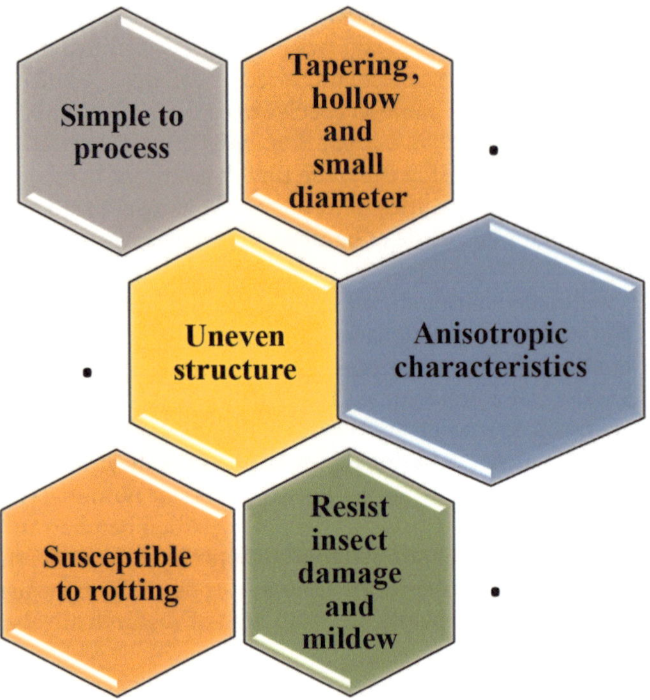

Figure 1.
Unique properties of bamboos.

Figure 2.
Different structural shapes of bamboo (adopted from open access). (https://www.guaduabamboo.com/blog/how-to-bend-bamboo and https://www.wikihow.com/Bend-Bamboo).

occasionally in conjunction with an inner core element. Similar to the square shape, a squared cross-section can be produced by compressing a bamboo stalk into a square. Bamboo may be bent into arches by controlling its growth to take on the desired shape. Using this method would have been less expensive than purchasing regular timber to achieve the same result. Additionally, traditional techniques like exerting pressure and heat are used to create the flatter and curved bamboo shapes. A stronger structural element will be created when you combine two or maybe more pole in the same curve. In many regions of South East Asia, this is by far the popular technique for curving bamboo utilised in big scale bamboo building.

2.2 Characteristics of bamboo and its fibre

Large-scale bamboo structures consist of a hollow section and a narrow wall with a particular tapered form [23]. There are over 10 cellular layers in bamboo, each with a unique alignment of the micro fibrils and alternately thick and thin layers on the cell wall of the bamboo fibres. Unidirectional bamboo fibres are strengthened by parenchyma cells ground tissue, which serves as a cellular matrix (**Figure 3**). An indication of bamboo's functionally graded materials is the unequal access of vascular tissue in the radii direction. The aforementioned structures and characteristics give outstanding bamboo durability, hardness, bending flexibility, and lightweight. Bamboo's primary chemical components include cellulose, hemicellulose, and lignin, as well as various extractions, a small amount of ash, and silicon dioxide.

A single bamboo fibre's microstructure features layered cell walls arranged in concentric circles as depicted in **Figure 4**. And the layers comprise a substantial cell wall, a restricted lumen, a few pits, and a small micro fibril angle.

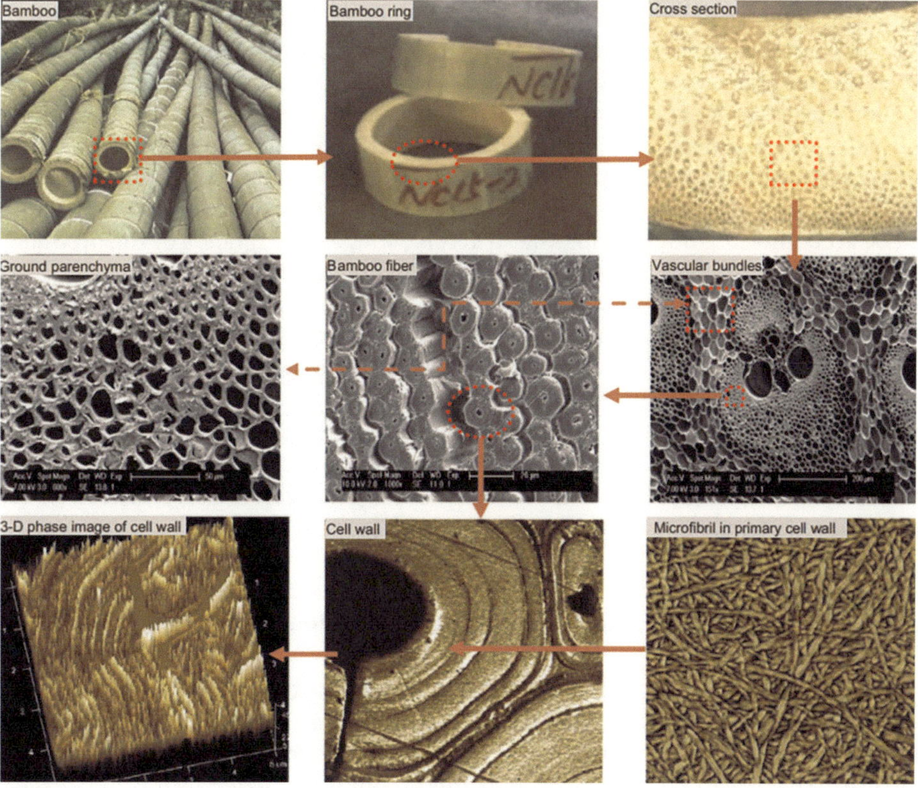

Figure 3.
Morphology and composition of bamboo (adapted from [24]).

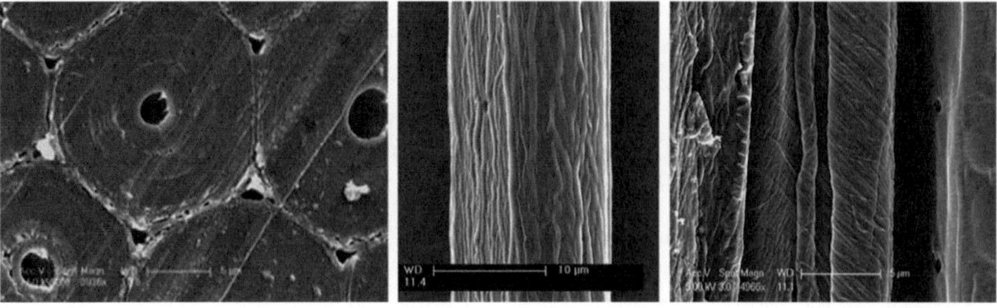

Figure 4.
Emission electron microscope picture of the longitudinal cross section of a single fibre (adapted from [24]).

3. Bamboo-based bamboo composites and panels

The production of bamboo-based composites and panels after being inspired by the advancements made in the woodworking sector was of great interest. Researchers created various types of bamboo-based composites due to their improved understanding of the unique characteristic of bamboo material and the bonding ability of bamboo green, medium, and yellow [25]. Comparing these novel products to wood-based products/panels reveals various distinguishing and attractive characteristics: The bamboo composites exhibit excellent proportions, dimensional stability, minimum distortion, and a constant size together with high wear resistance, adequate stiffness, and sufficient strength. Various other positive advantages such as low maintenance, high impact resistant, weather resistant, etc., can be witnessed in bamboo composite products. As reported in various papers, the bamboo fibres are treated with different polymeric resins like phenolformaldehyde, polyester, polylactic and Polypyrrole to develop a modified bamboo composite for structural application [26, 27]. At a specific amount of resistance to rot and insects, the customer's requests can be accommodated by altering the structure, size, and toughness of the product. Raw bamboo's improved qualities can be observed in many different contexts. To satisfy various needs, the product surfaces can be decorated in a number of ways.

Chen et al. [28] Making bamboo-bundle laminated veneer timber (BLVL) is a crucial step in producing long-span bamboo-based engineering materials. In the preprocessing densification process, phenol formaldehyde (PF) resin, polyvinyl acetate (PVAC), and other resins were combined to create the adhesive. These composites operate mechanically well, absorb moisture well, and bond strongly [29, 30] additionally, the BLVL exhibits outstanding bending characteristics, which allow for applications in bamboo structural engineering [31].

4. Structural applications of the advanced bamboo composites

Modifications and developments to the structural use of bamboo composite are being undertaken globally. Various researchers have researched and examined the use of bamboo composites developed using various resin materials for structural advances. The newest developments in bamboo composite materials for their structural applications undertaken by different scientists and researchers are presented here in this section.

Javadian et al., [32] reported on the unique usage of bamboo as sustainable alternative to synthetic fibres and studied their technical feasibility as reinforcement for structural-concrete beams for producing bamboo fibre-reinforced polymer composites. The bamboo used for this method was *Dendrocalamus asper*, popularly called as Petung Putih bamboo. The bamboo sections were first cooked at 80°C in normal water in a sealed jar for 8 to 20 hours. By reducing the lignin interface's adherence to the cellulose fibres, boiling the bamboo segments helped soften their microstructure, making bundling the segments into fibres simpler. The mix proportions are hand-laid up to generate them. Bundles of the coated, infused bamboo fibres were then placed into a hot-press machinery mould in the fibre direction after being initially dipped with an epoxy resin matrix. A simple yet effective semi-automatic hot-press compression moulded machine with high pressure of 25 MPa and a maximum temperature of 140°C was utilised. Eventually, specimens of bamboo composites were created using the authors' unique processing technology. The results of this study show that the ultimate loads for bamboo composite concrete beams and fibre reinforced polymer reinforced concrete beam that adhere to the ACI standard are comparable. ACI 440.1R-15, "Guide for the development and fabrication of structural concrete reinforced with Fiber Reinforced Polymer (FRP) bars," has provided design standards for the use of FRP materials as reinforcement in concrete. As opposed to steel reinforced concrete members, the FRP reinforced concrete has less ductility, which can be attributed to these guides. ACI 440.1R-15, that takes ductility into account, was the main and most applicable design guide to be employed in this research to analyse the effectiveness of bamboo composite reinforced concrete beams. The principles of equilibrium and compatibility serve as the cornerstone for the design recommendations provided by ACI 440.1R-15. It was decided to use the ultimate strength design method instead of the working stress conceptual design to generate the concrete beams supplemented with bamboo composite reinforcement in order to achieve results that were comparable to those obtained using methodologies used by other standards, such as ACI 318 for steel reinforced concrete design. The outcomes show the newly created bamboo composite materials potential for usage as a novel element for reinforced structural concrete beams. The bamboo composite reinforcement proposed in the paper has a lot of potential for use in real-world applications. This study also demonstrated that these newly created bamboo composite materials can be used to construct low-cost, low-rise housing units in situations where it is challenging to obtain steel reinforcement, there is little demand for ductility, and secondary-element failure adequately forewarns collapse. **Figure** 5 shows the longitudinal and transverse bamboo-composite reinforcement system created for this investigation. This study thus exhibits the newly created bamboo composite material's potential for usage as a novel element for reinforced structural concrete components.

A review was proposed by [18] on three categories of novel engineered wood composites, including cross-laminated timber (CLT), fibre-reinforced polymer (FRP) reinforced glulam, and timber scrimber. Also, three types of novel engineered bamboo composites, including glued-laminated bamboo (glubam), laminated bamboo lumber (LBL), and bamboo scrimber were reported.

Engineered wood satisfying the modern structural design suitable for structural applications has been developed in the last decades, embracing wood as one kind of dominant material for modern structures [12]. Similarly, engineered bamboo is gaining increasing interest in the current scenario from structural or construction industries, owing to its relatively low variability in material

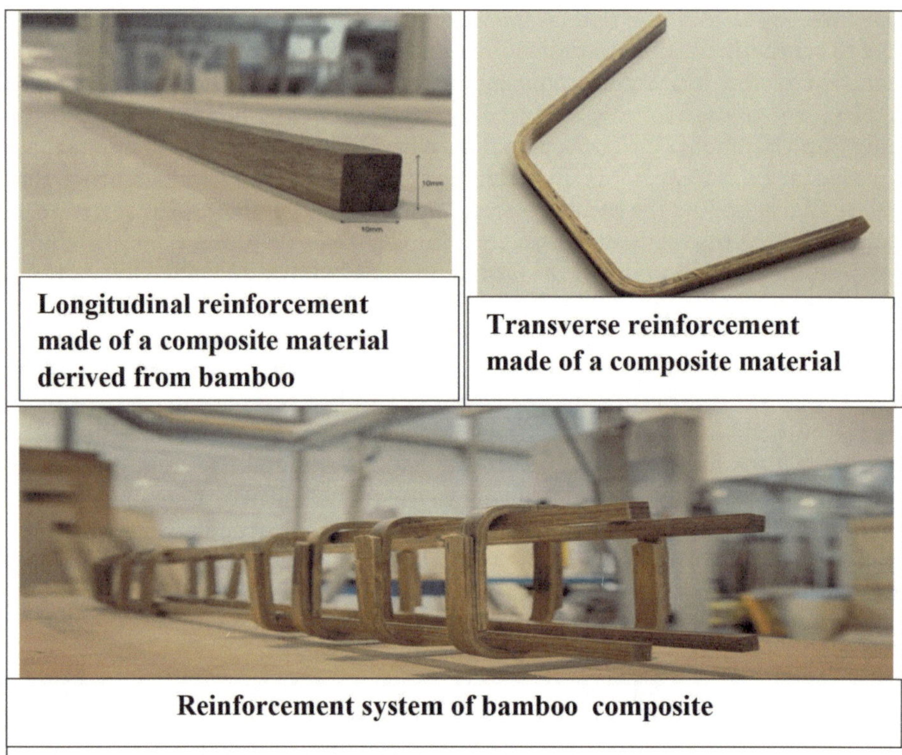

Figure 5.
Longitudinal and transverse reinforcement derived from a bamboo composite material (adapted from [32], with permission).

properties and shape standardisation. Three commonly used engineered bamboo products for structural applications primarily manufactured in China, and is used for the buildings or bridges construction are glued-laminated bamboo (glubam), laminated bamboo lumber (LBL), and parallel strand bamboo (PSB), which are described in detail below:

i. Laminated bamboo lumber (LBL)

Laminate bamboo lumber (LBL) is a common engineered bamboo product made from bamboo components that have been glued together in various ways (for example, into strands, strips, mats, etc.) to make boards with rectangle cross sections.

The following can be used to summarise one of the precise processes used to create bamboo-bundle laminated veneer lumber (BLVL) [33], which is basically one type of LBL products:

1. Each bamboo tube is divided into four pieces that are almost the same size, and the strips are broomed and rolled into laminated reticulate sheets using an untwining machine;

2. The bamboo sheets are sliced into bamboo bundles, which are then air-dried after they are cross-linked in the width direction.

Figure 6.
Systematic representation of LBL manufacturing process (adapted with permission from [18].

3. Using cotton thread, bamboo bundles of thickness 5–7 mm are chosen and positioned in the breadth direction before being knotted together to make a homogeneous one-piece veneer.

4. After submerging in phenol-formaldehyde (PF) and drying, the bamboo veneers are hot pressed into a single BLVL by means of the specially made mould, as shown in **Figure 6**.

Li et al. [34] highlighted a theoretical method for calculating the folding volume of LBL beams and presented a simplified strain-stress relationship for LBL. The internal joints observed in the LBL beams were discovered to have a more significant impact on the specimen during tangent bending than those undergoing radial bending.

ii. Glued laminated bamboo

The term "glued-laminated bamboo" and the trademark "glubam," which alludes to the concept of "glulam" were created for a ground-breaking engineered bamboo product. The five steps that make up the complete global manufacturing process (shown in **Figure 7**) are:

1. choosing raw bamboo;

2. dividing bamboo culms into strips with a thickness of 2–3 mm and a width of 30 mm, and removing wax from the culms' surfaces,

3. bamboo curtains forming by the arrangement of the bamboo strips in parallel to one another;

Figure 7.
Manufacturing process of Glubam (adapted from [18] with permission).

4. phenol-formaldehyde resin is used to glue the manufactured bamboo curtains,

5. followed by hot pressing the stacked bamboo curtains to make a single bamboo sheet. The final step is to finger-join these bamboo sheets to create structural elements.

iii. Parallel strand bamboo

Parallel strand bamboo (PSB), known as Bamboo scrimber is produced by compacting fibre bundles that have been crushed and soaked with resin into dense blocks with rectangular cross-sections. Approximately 80% of the raw resources can be used in the resource-efficient production method for PSBs products with multiple mechanical properties suitable for structural applications. Following is a summary of the complete PSB production process: (1) slashing bamboo culm into the bamboo two halves and expelling the bamboo notes; (2) straightening these bamboo halves into bundles of bamboo fibres; breaking in the longitudinal direction and connecting in transversal direction (3) drying those fibre bundles and soaking them in resin PF; and, finally, (4) hot-pressing these resin-soaked bundles of fibre into billets till the resin is cured making the desired PSB product.

Glubam has reasonable mechanical qualities and higher performance when related to other engineered bamboo products, but its relatively high density is a negative that needs to be addressed in the future. Additionally, PSB can offer extremely high parallel-to-grain, however due to its extremely high density, the dimensions of the

Figure 8.
A) Frame structure with LBL beams and PSB columns. B) Figure of LBL serpentine corridors (a) exterior (b) interior. C) Reinforced glubam beams at WWU (a) main gymnasium (b) natatorium (adapted with permission from [18]).

cross sections of the construction that uses PSB are constrained. However, China-based PSB engineering projects have already illustrated their suitability for structural applications. The application of these tailored bamboo composites for structural purposes is also covered in this review article as depicted in the **Figure 8**:

Although engineered bamboo is used in ornamental and surface applications worldwide, structural uses are only now starting to materialise. In both main and secondary structural applications, laminated bamboo and bamboo scrimber were used, as demonstrated by case studies provided by [35]. The chapter's case studies demonstrated industry advancement by using bamboo scrimber. To separate the fibres and keep the matrix, the raw, green bamboo culm is split and then crushed. The material is then impregnated by dipping the fibre bundles in a resin (such as phenol-formaldehyde). Then, the strips are placed in a mould and cold-pressed into the beam's structure after being soaked in resin. Finally, the composite beams are heated for curing and are ready for application.

These composite beams can be utilised for various structural applications:

a. Utilisation of bamboo scrims in urban construction:

   This work highlighted the utility of bamboo scrims in urban infrastructure construction, for traditional surface applications, particularly those where durability is a factor, bamboo scrimber have been used. The applied product's flexural strength is more vital than that of frequently used wood in these applications because of the product's extremely high density. The findings showed that the bamboo scrimber was appropriate for these uses; the flooring system could attain a stiffness that restricted deflection to situations in which

Figure 9.
Gare du Nord Station infrastructure by bamboo scrimber composite (adapted from [35] with permission).

sub-beams supported it. According to **Figure 9**, the flooring was laid in 2008 and the station has over 200 million visitors annually.

b. Utilisation of bamboo composite in structural sector:

This case illustrates how bamboo composite can be used structurally, an example of the material's potential. **Figure 10** shows the home in the Moso Bamboo Modern Technological Park in china.

Pozo Morales et al., [36] worked on developing a biodegradable composite for structural use as the primary objective of the work. This was accomplished by combining the suitable species of bamboo reinforcement with an appropriate natural thermoplastic polylactic acid (PLA) biopolymer, developing an inventive extraction procedure, and optimising the production process parameters. Bamboo strips were measured

Figure 10.
Bamboo house, Moso bamboo modern Technological Park, Anji, China (adapted from [35] with permission).

Figure 11.
Manufactured bamboo – PLA composite panels (adapted from [36]).

thickness of 1.5 mm and 1500 mm lengthwise. They were further machined into the necessary lay-up and form to create laminates with specified qualities stronger and stiffer than those of traditional E-Glass/Epoxy laminates, as depicted in **Figure 11**. The bamboo strips were joined together using a PLA matrix to meet the criteria for biodegradability. This study's novel mechanical extraction technique can recover natural strip reinforcements at high levels and low prices without harming the environment because no chemical treatments are applied. The discovered material's prospective aircraft uses are fairly limited. Still, its high mechanical qualities may meet the needs of many other manufacturing industries, such as the automobile or energy industries. Glass fibre composites are frequently used to make wind turbine blades, and the substance may also be used in various automotive surfaces. Both industries choose structural materials primarily based on price. Potential markets could also include housing and watercraft.

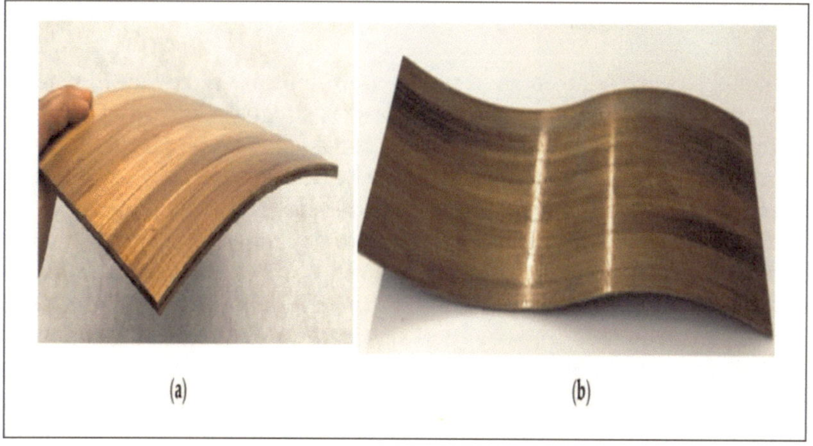

Figure 12.
Demonstrative bamboo panels from bamboo –PLA composite (a) cylinder concave shape (b) concave and convex shape (adapted from [36] with permission).

Two demonstration laminate components with curved geometries were created in order to demonstrate that it is possible to create structures with complex geometries, which is a crucial issue for applications in the future (see **Figure 12**).

4.1 Innovative applications of bamboo composite

Here are some creative ways that bamboo composites have been used. Here are a few instances of structural applications in several areas from a global perspective. Most users are aware of the benefits of bamboo products and support attempts to live sustainably in the world. This is further strengthened by the fact that it has received recognition for its excellence as an innovative material from a variety of sources, demonstrating that hybrid bamboo composite material can outperform other types of materials in a variety of areas including physical, mechanical, and aesthetic such as architecture, construction, interior design, and even auto parts Suhaily et al., [37]. **Table 1** highlights some innovative designs and applications of bamboo fibre bio-composites in various categories.

S.No	Category	Name of Product	Reference
1.	Interior Design	Basketball Court	Smith and Fong [38]
		Bamboo Dome	Nghia [39]
2.	Building and Construction	Kontum indochine café	Nghia [39]
3.	Furniture design	Infinity bench	Williams [40] and Huang [41]
		Hangzhou Bent Bamboo Stool	Min [42]
4.	Automotive components	Renault Megane Bioconcept car (Cloth seats, floor, dashboards)	Makinejad [43]

Table 1.
Innovative designs and applications of bamboo fibre bio-composites in various categories.

5. Conclusion

This chapter's primary objective is to concisely summarise advanced bamboo composites' potential, mainly for structural use. The technological and application advancements are highlighted in this chapter section. Many people are using bamboo as a "green" fibre. It is a natural fibre and may be grown quickly. To minimise any negative ecosystem devastation and manufacture reasonably priced polymeric reinforced composites, researchers are striving to develop composites using natural fibres that are completely biodegradable. As a result, there has been a noticeable rise in the use of bamboo fibres as composite reinforcement material recently. These materials are primarily used for structural applications in the housing and construction industries in the form of standard regular shapes like beams, planks, lumbers, and other elements, among other things. The introduction to advanced bamboo composites as engineered materials is covered in one of the chapter's sections. Additionally, it emphasises the critical notion of method and technology for bamboo composites and its numerous structural applications. The research and data of current studies on bamboo fibre-based composites describe multiple structural applications of bamboo composites. According to the data, examples of structural applications include the display panels of bamboo and PLA composite and the installation of lighter-weight bamboo-based walls in bamboo steel homes. This chapter has tried to enhance the knowledge about advanced bamboo composite for different structural applications. The vision of working towards sustainable approaches using bamboo as the natural raw material helps save the environment from the harmful effects of heat. Better architecting and designing the bamboo-based structural thermal insulating material can further boost the final product's versatile characteristics, especially in structural applications.

Acknowledgements

The authors thank Director CSIR-AMPRI, Bhopal, for providing the necessary institutional facilities and encouragement.

Conflicts of interest

The authors declare no conflict of interest.

Author details

Medha Mili[1,2], Anju Singhwane[1], Vaishnavi Hada[1], Ajay Naik[1,2], Prasanth Nair[1,2], Avanish Kumar Srivastava[1,2] and Sarika Verma[1,2*]

1 Council of Scientific and Industrial Research - Advanced Materials and Processes Research, Institute (AMPRI), Bhopal, M.P, India

2 Academy of Scientific and Innovative Research - Advanced Materials and Processes Research Institute (AMPRI), Bhopal, M.P, India

*Address all correspondence to: drsarikaverma20@gmail.com; drsarikaampri@gmail.com

References

[1] Mitch D, Harries KA, Sharma B. Characterization of splitting behavior of bamboo culm. Journal of Materials in Civil Engineering. 2010;**22**:1195-1199

[2] Valentine R, Matteucci G, Dolman AJ, Schulze E-D, Rebmann C, Moors EJ, et al. Respiration as the main determinant of carbon balance in European forests. Nature. 2000; **404**(6780):861-865

[3] Broadmeadow M, Matthews R. Forests, carbon and climate change: The UK contribution. Forest Communication Information Note. 2003;**48**:1-12

[4] Asdrubali F, Ferracuti B, Lombardi L, Guattari C, Evangelisti L, Grazieschi G. A review of structural, thermo-physical, acoustical, and environmental properties of wooden materials for building applications, build. Environment. 2017; **114**:307-332

[5] Hillis WE. Heartwood and Tree Exudates. Volume 4 of Springer Series in Wood Science. Berlin, Heidelberg: Springer; 1987

[6] Gupta A, Kumar A, Patnaik A, Biswas S. Effect of different parameters on mechanical and erosion wear behavior of bamboo fiber reinforced epoxy composites. International Journal of Polymer Science. 2011;**2011**:1-11

[7] Mili M, Hashmi SAR, Tilwari A, et al. Preparation of nanolignin rich fraction from bamboo stem via green technology: Assessment of its antioxidant, antibacterial and UV blocking properties. Environmental Technology. 2023;**44**(3):416-430. DOI: 10.1080/09593330.2021.1973574

[8] Verma S, Hashmi SAR, Mili M, Hada V, Prashant N, Naik A, et al. Extraction and applications of lignin from bamboo: A critical review. European Journal of Wood and Wood Products. 2021;**79**(6):1341-1357

[9] Mili M, Verma S, Hashmi SAR, Gupta RK, Naik A, Rathore SKS, et al. Development of advanced bamboo stem derived chemically designed material. Journal of Polymer Research. 2021b; **28**(5):1-11

[10] Xiao FM, Fan SH, Wang SL, Xiong CY, Zhang C, Liu SP. Carbon storage and spatial distribution in Phyllostachy pubescens and Cunninghamia lanceolata plantation ecosystem. Acta Ecologica Sinica. 2007; **27**(7):2794-2801

[11] Yiping L et al. Technical Report 32: Bamboo and Climate Change Mitigation. Beijing, China: INBAR; 2010

[12] Li Z, Zhou R, He M, Sun X. Modern timber construction technology and engineering applications in China. Proceedings of Institution of Civil Engineers Civil Engineering. 2019; **172**(5):17-27

[13] Reis JP, de Moura MFSF, Silva FGA, Dourado N. Dimensional optimization of carbon-epoxy bars for reinforcement of wood beams. Composites Part B-Engineering. 2018;**139**:163-170

[14] Lu T, Jiang M, Jiang Z, Hui D, Wang Z, Zhou Z. Effect of surface modifi cation of bamboo cellulose fi bres on mechanical properties of cellulose/epoxy composites. Composites Part B: Engineering. 2013;**51**:28-34

[15] Liu X, Smith GD, Jiang Z, Bock MCD, Boeck F, Frith O, et al. Nomenclature for engineered bamboo. BioResources. 2016;**11**(1):1141-1161

[16] Wang G, Yu Y, Shi SQ, Wang JW, Cao SP, Cheng HT. Microtension test method for measuring tensile properties of individual cellulosic fibers. Wood and Fiber Science. 2011;**43**(3):251-256

[17] Liese W. Research on bamboo. Wood Science and Technology. 1987;**21**(3): 189-209

[18] Sun X, He M, Li Z. Novel engineered wood and bamboo composites for structural applications: State-of-art of manufacturing technology and mechanical performance evaluation. Construction and Building Materials. 2020;**249**:118751

[19] Mili M, Hashmi SAR, Ather M, Hada V, Markandeya N, Kamble S, et al. Novel lignin as natural-biodegradable binder for various sectors—A review. Journal of Applied Polymer Science. 2022;**139**(15):51951

[20] Ramage MH, Burridge H, Busse-Wicher M, et al. The wood from the trees: The use of timber in connection. Renewable and Sustainable Energy Reviews. 2017a;**68**:333-359

[21] Ramage M, Sharma B, Shah D, Reynolds T. Thermal relaxation of laminated bamboo for folded shells. Materials and Design. 2017;**132**:582-589. DOI: 10.1016/j.matdes.2017.07.035

[22] Zhong Y, Wu GF, Ren HQ, Jiang ZH. Bending properties evaluation of newly designed reinforced bamboo scrimber composite beams. Construction and Building Materials. 2017;**143**:61-70

[23] Liu DG, Song JW, Anderson DP, Chang PR, Hua Y. Bamboo fiber and its reinforced composites: Structure and properties. Cellulose. 2012;**19**(5): 1449-1480

[24] Wang G, Chen F. Development of bamboo fiber-based composites. Advanced High Strength Natural Fibre Composites in Construction. 2017: 235-255

[25] Qisheng Z, Shenxue J, Yongyu T. Industrial Utilization on Bamboo. Beijing, China: International network for bamboo and rattan; 2002

[26] Khalil HA, Alwani MS, Islam MN, Suhaily SS, Dungani R, H'ng YM, et al. The use of bamboo fibres as reinforcements in composites. In: Biofiber Reinforcements in Composite Materials. Sawston, Cambridge: Woodhead Publishing; 2015. pp. 488-524

[27] Zhang K, Chen Z, Smith LM, Hong G, Song W, Zhang S. Polypyrrole-modified bamboo fiber/polylactic acid with enhanced mechanical, the antistatic properties and thermal stability. Industrial Crops and Products. 2021;**162**: 113227

[28] Chen F, Deng J, Cheng H, et al. Impact properties of bamboo bundle laminated veneer lumber by preprocessing densifification technology. Journal of Wood Science. 2014;**60**(6):421-427

[29] Zhang D, Wang G, Ren W. Effect of different veneer-joint forms and allocations on mechanical properties of bamboo-bundle laminated veneer lumber. BioResources. 2014;**9**(2): 2689-2695

[30] Deng J, Chen F, Li H, Wang G, Shi SQ. The effect of PF/PVAC weight ratio and ambient temperature on moisture absorption performance of bamboobundle laminated veneer lumber. Polymer Composites. 2016; **37**(3):955-962

[31] Chen F, Jiang Z, Wang G, Li H, Simth LM, Shi SQ. The bending properties of bamboo bundle laminated

veneer lumber (BLVL) double beams. Construction and Building Materials. 2016;**119**:145-151

[32] Javadian A, Smith IFC, Hebel DE. Application of sustainable bamboo-based composite reinforcement in structural-concrete beams: Design and evaluation. Materials. 2020; **13**(3):696

[33] Zhou H, Sun F, Li H, Zhang W, Cheng H, Chen L, et al. Development and application of modular Bamboo-Composite Wall construction. BioResources. 2019;**14**(3): 7169-7181

[34] Li G, Wu Q, Zhang AJ, Deeks J. Su, ultimate bending capacity evaluation of laminated bamboo lumber beams. Construction and Building Materials. 2018;**160**:365-375

[35] Sharma B, van der Vegte A. Engineered bamboo for structural applications. In: Nonconventional and Vernacular Construction Materials. Woodhead Publishing; 2020. pp. 597-623

[36] Pozo Morales A, Güemes A, Fernandez-Lopez A, Carcelen Valero V, De La Rosa Llano S. Bamboo–polylactic acid (PLA) composite material for structural applications. Materials. 2017; **10**(11):1286

[37] Suhaily SS et al. Bamboo Based Biocomposites Material, Design and Applications. 2013

[38] Smith & Fong. PlybooSport Bamboo Sport Flooring [Online]. San Francisco: Smith & Fong Company; 2010. Available at http://www.plyboo.com/

[39] Nghia VT. Vo Trong Nghia Architects [Online]. Vietnam. 2010. Available at http://www.votrongng hia.com

[40] Williams A. Infinity Bamboo Bench [Online]. New York. 2010. Available at http://www.designaw.me

[41] Huang T. TomHuangStudio.Com [online]. 2007. Available at http://www.tomhuangstudio.com.

[42] Chen MA. Hangzhou Bamboo Stool [Online]. China. 2013. Available at http://chen-min.com

[43] Makinejad MD, Salit MS, Ahmad D, Ali A, Abdan K. A review of natural fibre composites in the automotive industry. Research on Natural Fibre Reinforced Polymer Composites. 2009;**16**: 247-262